T0219716

Emergency Characterization of Unknown Materials

Emergency Characterization of Unknown Materials
Second Edition

Rick Houghton and William Bennett

CRC Press
Taylor & Francis Group
Boca Raton London New York

CRC Press is an imprint of the
Taylor & Francis Group, an **informa** business

Second edition published 2021
by CRC Press
6000 Broken Sound Parkway NW, Suite 300, Boca Raton, FL 33487-2742

and by CRC Press
2 Park Square, Milton Park, Abingdon, Oxon, OX14 4RN

First edition published by CRC Press 2007

CRC Press is an imprint of Taylor & Francis Group, LLC

ISBN: 978-0-367-48025-7 (hbk)
ISBN: 978-1-003-03766-8 (ebk)

Typeset in Minion
by Deanta Global Publishing Services, Chennai, India

Contents

Contents vii

3   Detection Technology   121

Contents

Acknowledgments

Many people have contributed to this book in many ways, because no one can claim to be an expert on everything. It takes a team.

Warfighters, first responders, manufacturers, trainers and others have a need to identify or characterize hazardous materials to complete their mission. We hope this book helps.

We would like to simply say thank you to those who keep us safe.

Authors

Rick Houghton recently retired after 40 years as a first responder and trainer. His career began as a firefighter, paramedic and hazardous materials trainer. He has extensive experience as a professional trainer and course developer.

William Bennett is a professional trainer, course developer and business owner. He recently retired from the U.S. Air Force with 35 years' experience, including 16 years as a founding member of the 51st Civil Support Team.

Introduction

It turns out that the question is much easier than the answer. When you ask, "What is that stuff?" the answer depends on your job and the situation in which the material is found. Some jobs require a high degree of precision; other jobs require a sweeping overview of the hazards. This book is written for the latter situation and is intended for use by workers who are responsible for a safe response to and management of unknown hazardous materials. The book is written with an emphasis on public safety and the management of life safety hazards. You will find technology and strategies to identify the hazards presented by an unknown material. You will not necessarily find specific tests to identify a material by name, although some technology is able to do so in some cases. The test methods presented within involve manipulation of small amounts of sample – literally a hands-on approach. The theme of this book is the emergency characterization of unknown material.

Risk analysis is a procedure to identify threats and vulnerabilities, to analyze them to ascertain the exposures, and to highlight how the impact can be eliminated or reduced. If an unknown material can be identified with a high degree of confidence, many resources become available for risk analysis. However, identification of an unknown material before risk analysis is not a prerequisite and, indeed, might not be possible. The risk analysis that you develop will be a fluid concept that changes with the circumstances of any situation. Risk analysis based on hazard characteristics of an unknown material is a sound practice, especially during an emergency.

The hazard identification methods presented in this book are intended for use by frontline workers. Firefighters and hazardous materials teams, when responding to a hazardous material emergency involving an unknown substance, are primarily interested in protecting public safety. Identifying the hazards of the unknown substance is essential to plan development and sustained actions.

Law enforcement officers, particularly those involved with drug enforcement and bomb squads, are often faced with unlabeled or mislabeled chemicals, usually in mixtures, in clandestine labs. Special tactical teams may be required to support those who would render safe an improvised hazardous material device intended to cause harm. Crime scene investigators may need to work in a hot zone to identify, collect and preserve evidence that is complicated by possible contamination with hazardous material.

Waste site workers regularly encounter unknown material in containers with labels that have long since vanished. Community waste collections produce batches of unlabeled containers clustered in cardboard boxes in car trunks. Characterizing hazards of these materials before bulk packaging and transport will decrease risk.

Lab workers often have a need for a rapid triage procedure when processing incoming material. This includes centralized facilities that may become chemical triage facilities during periods of national emergency as described in a government plan.

As you can see, there is a need for straightforward, concise and practical guidance in many working environments. Anyone with experience in emergency response knows the importance of working with a plan and the need for guidance that is easy to remember. Your response will be more effective if you develop your plan ahead of time rather than accepting spontaneous suggestions from several well-intentioned people.

This book is inclusive of as many field workers as possible, regardless of their skill level. It opens with a chapter that provides a basic overview of approximately 50 chemical and physical terms and definitions listed in National Fire Protection Association 472 as hazardous material technician-level competencies. NFPA is a long-respected champion of recommended competencies.

The next chapter focuses on types of hazards presented by chemical compounds and mixtures, organisms and radiation sources. Identifying these hazards is the purpose of this book. Detection technology for the purpose of identifying hazards is reviewed in the third chapter. General technological advantages and disadvantages are examined relative to the identification of hazards.

The fourth chapter covers strategies for the identification of hazards. This chapter mentions tactical considerations but is not all-inclusive. The site supervisor or incident commander is responsible for developing a work plan to resolve the situation once the hazards have been identified. This strategy chapter includes incidents involving weapons of mass destruction (WMD). As you will see, suspicion of a WMD, illegal drug, or explosive material will alter the strategy for identifying hazards.

New to this second edition is a fifth chapter on techniques. Raman, FTIR and HPMS have become the dominant technologies in the field, used to identify unknown material quickly. This techniques chapter describes several methods of manipulating samples or taking an extra step to get the most out of these wonderful instruments.

The scope of this book does not include sampling techniques beyond a quick overview of robotic sampling. Sampling is contingent on the unique needs of a responding agency and is too wieldy a topic to include here. Law enforcement has distinct requirements for evidence collection, while a hazardous materials team requires a sampling technique that is rapid and safe. There are sampling resources available based on your individual need.

References are included with each chapter, and a comprehensive index ends the publication.

Helping you answer that question, "What is that stuff?" becomes overwhelming without defining the scope of the book. There is always more information that could make the answer more specific, but the cost is reflected in a resource too bulky and overwhelming for convenient use. While compiling this book, we were constantly aware of the need to segregate the material by that which could be reasonably remembered and utilized by workers in the field or lab. There is a lot of amazing science out there that is very interesting but not necessarily applicable to your job. Filtering that information for you is the main purpose of this book.

Our hope is you will take what is helpful from this book and use it. Employers and employees should partner to provide the safest working conditions possible. You can use the information in this book to improve workplace health and safety. You can also use this resource to protect life, stabilize an emergency situation and reduce property loss, while improving your response to unknown substances.

Terms and Definitions 1

Figure 1.1 Common terminology and an understanding of definitions keep everyone on the same page.

This chapter reviews the terms and definitions that should be familiar to hazardous material technicians based on recommendations by the National Fire Protection Association (NFPA) and others. These terms are part of the NFPA competencies recommended for emergency responders who must respond to a hazardous material threat and must possess the knowledge and skills needed to safely perform tasks related to the response (Figure 1.1). These recommendations are pertinent to other workers, such as hazardous waste technicians, laboratory technicians, and anyone in need of a rapid laboratory triage process. You should be able to describe the terms listed in Table 1.1 and explain their significance in risk assessment.

Table 1.1   NFPA 472 Technician Competencies: Hazardous Materials Terms and Definitions*

Corrosive (acids and bases/alkaline)
Air reactivity
Auto refrigeration
Biological agents and biological toxins
Blood agents
Boiling point
Catalyst
Chemical change
Chemical interactions
Compound, mixture
Concentration
Critical temperature and pressure
Dissociation (acid/base)
Dose
Dose response
Expansion ratio
Fire point
Flammable (explosive) range (LEL and UEL)
Flashpoint
Half-life
Halogenated hydrocarbon
Ignition (auto ignition) temperature
Inhibitor
Instability
Ionic and covalent compounds
Irritants (riot control agents)
Maximum safe storage temperature (MSST)
Melting point and freezing point
Miscibility
Nerve agents
Organic and inorganic
Oxidation potential
Persistence
pH
Physical change
Physical state (solid, liquid, gas)
Polymerization
Radioactivity
Reactivity
Riot control agents
Saturated, unsaturated (straight and branched), and aromatic hydrocarbons
Self-accelerating decomposition temperature (SADT)
Solubility
Solution and slurry
Specific gravity
Strength
Sublimation

(Continued)

Table 1.1 (Continued)   NFPA 472 Technician Competencies: Hazardous Materials Terms and Definitions*
Temperature of product
Toxic products of combustion
Vapor density
Vapor pressure
Vesicants (blister agents)
Viscosity
Volatility

*Presentation in text has been altered from NFPA's alphabetical listing. See the table of contents for text presentation.

The theme of this book is identification of hazardous characteristics present in an unknown material so that you may respond more safely to a threat. This chapter is an overview based on relevance to your needs in the workplace. Utilization of safe work practices while identifying hazards will protect you and others from these yet unidentified hazards. In order to work safely, you must have the knowledge and skills necessary to do so.

This chapter is divided into three main sections; physical terms, chemical terms, and then weapons of mass destruction (WMD) terms. These definitions will help you to understand the terms both individually and as they relate to each other for the purpose of identifying and understanding hazardous characteristics of unknown substances. The intent is not necessarily to rigidly define each as a physical, chemical, or WMD term. All of the NFPA terms and definitions are covered in this chapter, but the material has been altered from the alphabetical list in order to clarify the presentation (Table 1.1).

Physical Terms and Definitions

Physical State

The physical state of the unknown substance you may characterize describes its phase as solid, liquid, or gas. There are other phases, such as plasma, but they are not likely to be encountered while identifying hazardous characteristics of a sample of unknown material.

A material may exist in one, two or all three phases simultaneously. Temperature and pressure influence pure substances in this manner, but mixtures or contaminants also affect phase. An ice cube floating in a glass of water shows water can exist in liquid and solid phase simultaneously based on heat content. Water can also exist in solid and liquid phase below its freezing point if salt is poured on the ice.

Solids

Solids tend to be the least hazardous phase because a solid is not very mobile. If the solid is finely divided, the solid may become more mobile as air currents move it. The smaller the particle, the more mobile the solid. Anthrax spores, asbestos and fiberglass fibers, silica and talc dust, and smoke particulates are just a few examples of small, mobile solid particles.

These small particles are measured in the scale of microns (one-millionth of a meter). A human hair can range from about 20 to 180 microns in diameter. Unaided visual resolution of 1–5 microns might be possible, but don't count on it, even if lighting and contrast conditions are ideal. Particles of one micron or less are easily inhaled deep into the alveoli of the lungs, where they are barely divided from the bloodstream. Particles of approximately 10 microns lodge in the bronchioles, which can stimulate asthmatic symptoms. Sometimes a liquid is adsorbed onto a small particle to change its intended application, such as a liquid pesticide in powder form.

Terms that describe solids are *particle, dust, powder, fume, particulate, fiber, flake, granule* and *soot*.

Liquids

Liquids are more mobile than most solids. Obviously, they assume the shape of the container and will seek the lowest level. Uncontrolled liquids will displace objects and soak into some solids. Liquids will also evaporate. If below its boiling point, liquid exists in the air as a vapor, which is several molecules clinging together and moving through the air.

Terms that describe liquids are *vapor, droplet, mist, fog, aqueous, viscous, solvent* and *volatile*.

Gases

Gases are the most mobile of the three phases. Gases exist as individual molecules and do not cling to each other as vapors do. Gases are obviously inhalation hazards, but are the easiest to dissipate in air. Gases can simultaneously exist as liquid under pressure, which is called "liquefied gas" – for example, liquefied petroleum gas.

Terms that describe gases are *liquefied gas, gaseous, compressed* and *cryogenic*.

Compound and Mixture

"Compound" is a loosely defined term meaning a single chemical substance – for example, sodium chloride or table salt. Individual elements combine to form a new compound that acts differently than the original elements. A compound consists of two or more elements, exists as a pure substance at its smallest physical division, and has characteristics different than those of the individual elements that form the compound. For example, both sodium and chlorine can exist as individual elemental compounds, but both are very reactive and unstable.

Mixture means a physical mixing of compounds to form a new product. A mixture consists of a diverse combination of compounds. For example, black powder is a physical mixture of sulfur, charcoal and potassium nitrate. Although the mixture may be finely ground, the three substances could be physically separated. They also can exist together in a container because there is no chemical reaction between them.

A hazardous material is often referred to as a compound. Mixtures of compounds may induce chemical reactions that produce new compounds with characteristics different than those of the original compounds.

After the black powder mentioned above is fired, the elements that made up the three diverse compounds recombine into new compounds, all of which are mixed to form the products of combustion, or smoke.

It is significant to note that reference material will provide data for pure compounds. A first responder cannot necessarily use reference data for individual hazardous materials to precisely predict the characteristics of a mixture of hazardous material. Impure material and contaminants should be expected with uncontrolled release of an unknown material. With some skill, broad predictions of hazardous characteristics are possible and practical.

Since exact reference data for real-world hazardous material releases is not possible, the first responder or lab technician may be able to glean as much or more information about the substance's characteristics by observation and direct manipulation of the material than would be available in the form of printed reference data.

Solution and Slurry

A solution is a physical mixing of compounds into a homogenous mixture. The compounds may be solid, liquid, or gas. The most common type of solution is formed when individual units of a compound (solute) are able to dissolve into a liquid (solvent) and stay suspended. A solution will not settle out but may be physically separated. An example is sugar dissolved in water. The sugar will not settle out of the water over time, but if the water is allowed to evaporate, the sugar will remain.

A slurry is formed when individual units of a compound are able to suspend in a liquid but they cannot stay suspended without continuing agitation. A slurry will settle out like mud in water. Examples of slurries include wet concrete, corn starch in water, and wet drywall compound. Water is often used to suspend a finely divided, nonsoluble compound as a slurry.

Knowing whether a hazardous material will form a solution or a slurry would help a first responder predict the characteristics of runoff water to a stream or storm drain.

Miscibility

"Miscible" means two or more compounds mix in all proportions. Miscibility describes the extent of the "mix-ability" of two compounds without limit.

For example, ethanol is miscible in water in all proportions. One gallon of ethanol could be added to one gallon of water, 100 gallons of water, or a swimming pool full of water. In each case, the ethanol would disperse evenly throughout the water. While the concentration would differ, all of the solutions would have uniform distribution of ethanol throughout the water. It would be impossible to saturate the solution so that excess ethanol remained undissolved in the water.

Note that both compounds are miscible in each other. That is, the same principle applies if one gallon of water is added to one gallon of ethanol, 100 gallons of ethanol, or a swimming pool full of ethanol.

Recognizing the characteristic of miscibility of an unknown substance in water or other known materials will help you predict the effects of runoff from hose streams into soil, storm drains, or sanitary sewers.

Miscible is applied to liquids and sometimes to gases. Miscible does not apply to solids. Immiscible means the two compounds don't mix to a significant degree.

Solubility

Solubility is the ability of a solid, liquid, or gas to dissolve into a solvent, which is typically a liquid. Water solubility means water is the solvent and another compound (the solute) is dissolved into it.

Solubility is usually expressed in terms of weight, volume, or percent. You might see this labeled as w/w, v/v, or w/v, where "w" is weight and "v" is volume. For example, the solubility of carbon dioxide is about 90 cm^3 of CO_2 per 100 ml of water. Solubility is affected by temperature and pressure.

When the maximum amount of solute is dissolved in the solvent, the solution is saturated. When directed to mix a saturated solution, solute is added to excess. For example, to form a saturated solution of sodium bicarbonate as a decontamination solution, simply add bicarb to water while stirring. When a little excess bicarb remains at the bottom of the container, the solution is saturated and cannot hold more at the current temperature.

During the production of soda, carbon dioxide gas is forced into water and the two substances exist as a liquid. Changes to the liquid in the form of temperature, pressure, pH, physical agitation, and others may initiate escape of carbon dioxide gas from the liquid, as seen in Figure 1.2. These same changes affect any gas dissolved in a liquid. Many liquid solutions around us contain dissolved gas, such as oxygen dissolved in lake water (fish need to breathe) and ammonia gas dissolved in water (glass cleaner).

Methyl ethyl ketone (MEK) is about 30% soluble in water. If equal volumes of water and MEK are mixed, about 30% of the MEK will dissolve into the water and some of the water will dissolve into the MEK. After agitation, equilibrium is established and about 70% of the MEK will float on the water while the other 30% mixes in the water layer (Figure 1.3).

The inverse also occurs, that is, water also enters the upper, MEK layer.

Adding more water will move more MEK from the upper layer into the lower water layer as well as more water into the upper layer. If the same volume of MEK is added to a volume of water twice as large as the volume of MEK, more of the MEK will dissolve into the water with less MEK available to float near the surface. This principle has a significant application in fire prevention and suppression. The solubility of some common compounds in different solvents appears in Table 1.2.

"Insoluble" describes the inability of a substance to form a solution (Figure 1.4). Motor oil is insoluble in water but soluble in paint thinner. Many substances are slightly soluble in practically any solvent, and the term is often used in a way that is not absolute. An insoluble solution might also be described in a safety data sheet or other source under a heading of solubility as negligible, slightly, or not soluble. You might be able to make a field estimate of solubility with accuracy of +/–10% and use that information to predict events.

Figure 1.2 Carbon dioxide effervescing from soda.

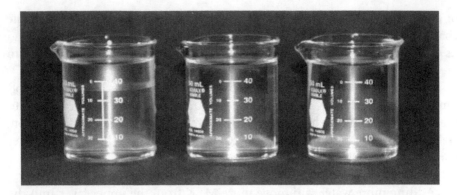

Figure 1.3 Methyl ethyl ketone (MEK) mixed with water at ratios of 50, 33, and 25% v/v (left to right). A relatively higher concentration of water in the solution produces a less obvious division between layers. The 50% solution produces a distinct separation of upper and lower layers in the solution (left). The 33% solution contains twice as much water compared to the volume of MEK and produces a thinner, less distinct layer at the top of the liquid (center). The 25% solution contains three times as much water as MEK and produces a very thin, nearly indistinguishable layer at the top of the liquid (right).

Table 1.2 Solubility of Common Compounds

Compound Name and Uses	Water	Ethanol	Ether	Acetone	Benzene
Methyl ethyl ketone (MEK) – PVC pipe primer	Soluble	Miscible	Miscible	Miscible	Miscible
Acetone – fingernail polish remover	Miscible	Miscible	Miscible	Miscible	Miscible
Isopropyl alcohol – rubbing alcohol	Miscible	Miscible	Miscible	Soluble	Very soluble
TNT – explosive	Insoluble	Soluble	Soluble	Soluble	Very soluble
Ethylene glycol – mixed with water for engine coolant	Miscible	Miscible	Soluble	Miscible	Slightly soluble

Something that is 100% soluble is not the same as saying it is miscible. While miscible solutions are mixable in all proportions, soluble solutions may be saturated. If a substance is 100% soluble, it means an equal amount is soluble in the solvent, but more cannot be dissolved.

Sugar, as used in baking, is sucrose and is highly soluble, but not miscible. The solubility of sucrose is about 200 g/100 ml. If sucrose is added to water while stirring, you will be able to add 200 g to 100 ml of water and get it all to dissolve. If you add more, it will sink to the bottom and will not be able to dissolve unless more water is added.

Emulsions are an apparent exception to insoluble compounds. Emulsions are small droplets suspended in a liquid. Very light oils will form an emulsion

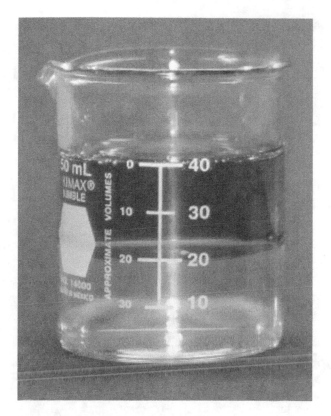

Figure 1.4 Insoluble oil floating on water. Insoluble materials separate after agitation and cannot be diluted.

in water after agitation. Emulsions often form milky or opaque liquid from clear liquids. Pesticides are sometimes dissolved in light oil, which is then mixed with water for application. The mixed solution is milky. Emulsions are not technically a solution, but may be designed to act like a solution for a specific application.

A solution is said to be saturated if no more solute can be dissolved into the solvent. Any solution that is not miscible can be saturated, such as the copper sulfate in water shown in Figure 1.5. Having a general understanding of the saturation point of a hazardous material in water will help you determine the magnitude of a spill in confined bodies of water or flowing water. Likewise, a hazardous material spilling into a body of water does not instantly mix and equalize the concentration throughout the water. A substance draining into a lake could easily produce a saturated solution locally with diminishing concentration at greater distance from the spill. If the lake remained calm, the solution would eventually reach equilibrium, but it could take a long time to do so.

Figure 1.5 Saturated solution of copper sulfate. Copper sulfate crystals dissolve the solid into the liquid while stirring. When the liquid becomes saturated, it can accept no more copper sulfate. Adding more solid only causes the excess crystals to settle to the bottom of the container. The addition of more water would allow the excess crystals to dissolve and become mobile.

A case study involved approximately 48,000 gallons of sodium hydroxide, also known as caustic soda or lye, that flowed into a creek when 28 Norfolk Southern railroad cars derailed on June 30, 2006, near the town of Gardeau, Pennsylvania (Figure 1.6). The spill damaged a highly productive trout stream and adjacent waterways. As a result of the spill, dead fish were reported up to 35 miles from the original accident site. The damage from the chemical spill was highly concentrated at the spill site and then dissipated through reaction and dilution as it washed downstream. All types of fish, from top-level predators through smaller fish and other aquatic life, were affected by the spill.

Given the volume and nature of the spilled chemical, it's likely all fish in Sinnemahoning Portage Creek downstream of the spill site were killed. Four miles of that stream were considered Class A Wild Trout Waters. The stream flows into the 20-mile-long Driftwood Branch and then eventually into Sinnemahoning Creek below the town of Driftwood in Cameron County.

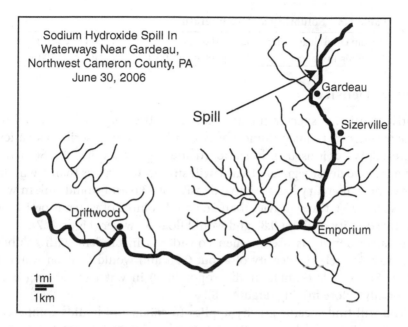

Figure 1.6 Map of Gardeau, Pennsylvania, and surrounding area. A 48,000-gallon sodium hydroxide release killed all aquatic life between the spill site and Emporium. Some insect life survived between Emporium and Driftwood and fish kill extended to Driftwood.

Warnings were issued to keep people, pets, and animals away from the water and stream banks.

Preliminary results of biological assessments showed essentially no living aquatic insects downstream from the spill site for approximately 7.5 miles. Samples collected in Emporium did show living organisms, but the stream was significantly impaired compared to normal conditions. Assessments completed below the Driftwood Branch found the numbers of aquatic insects to be relatively high. Although the sodium hydroxide dissipated relatively quickly, complete biological recovery of the stream is expected to take years.

Solubility is determined by the unique characteristics of the solvent and the solute as well as other variables, such as temperature and the presence of other compounds. Increased temperature does not necessarily mean you should expect increased solubility. Heating water will allow you to dissolve more table salt (sodium chloride) into the water than when the water is at a cooler temperature. However, water cannot hold as much ammonia gas when hot compared to colder solutions. In this case, the increased temperature works to drive the gas out of the solution (Table 1.3).

Solubility of solids in water will increase with increasing temperature. Solubility of gases in water will increase with decreasing temperature and increasing pressure. Increased heat will also drive higher vapor pressure (organic) liquids out of water solutions.

Table 1.3   Solubility and Temperature

Sodium chloride in water	35.7 g/100 cc @ 0°C	89.9 g/100 cc @ 0°C
Ammonia in water	39.1 g/100 cc @ 100°C	7.4 g/100 cc @ 100°C

Relative Density

Relative density, or specific gravity, is a ratio of the weight of a solid or liquid to the weight of an equal volume of water and is used to describe buoyancy of a material. Water is used as the standard and is given a value of 1. Something with a relative density greater than 1 will sink in water; less than 1 will float. For example, mercury has a relative density of 13.6 and would sink in water, because it is 13.6 times heavier than an equal volume of water. Motor oil has a relative density of about 0.87 and would float on water (Figure 1.7).

Relative density is useful when considered in tandem with solubility. Acetone has a relative density of about 0.78 and would float on water, but acetone is miscible (soluble in all proportions) in water so floating occurs only briefly before mixing (Figure 1.8).

You will find specific gravity applies to solids and liquids compared to water. This is useful because these hazardous materials most commonly come in contact with water. Most solid organic materials are lighter than water, with the exception being most halogenated hydrocarbons (Figure 1.9). Most solid inorganic material will sink in water.

Most organic liquids are lighter than water, with the exceptions again being those that are halogenated. There are only a few inorganic liquids and they possess unusual characteristics. Determining water solubility and specific gravity of an unknown material will help you eliminate or confirm many hazardous materials.

Relative density is a broader term than specific gravity, applying to liquids other than water and gases.

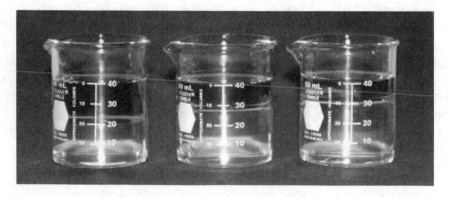

Figure 1.7 Motor oil mixed with water at ratios of 50, 33, and 25% v/v (left to right).

Figure 1.8 Water added to acetone. The agitation of a stream of water into acetone is enough to produce instant mixing and prevents the formation of a lighter layer of acetone at the top of the liquid. Gentle addition of acetone onto water would allow a layer of acetone to float, but its miscible characteristic would cause the layers to blend into a single, homogenous solution.

Viscosity

Viscosity is a term used to describe the "thickness" of a liquid or its resistance to flow. High viscosity liquids are thick and slow to pour. Viscosity is most often measured in a unit called "poise."

The term "poise" is in honor of Jean Louis Marie Poiseuille, a French physician and physiologist from the early- and mid-1800s. Specifically, poise is the metric unit of dynamic viscosity equal to one dyne-second per square centimeter. The corresponding International System of Units (SI) term is "pascal-second" or Pas. These are large units and references most often express viscosity in centipoise (cP) or millipascal-second (1/1000 pascal-second). Fortunately, 1 centipoise equals 1 millipascal-second. There are two ways to measure viscosity, and results are often described in terms of

Figure 1.9 Trichloroethylene (TCE) under water. TCE is insoluble in water and even after agitation will separate and fall below the water to form a distinct layer.

Table 1.4   Measured Viscosity of Common Substances

Substance at Room Temperature	Viscosity (cSt)
Brake fluid	12.7
Automotive antifreeze	17.6
Vegetable oil	65
Automatic transmission fluid	78
Motor oil, SAE 30	270
Corn syrup	5180

dynamic viscosity, measured in units of centipoise, and kinetic viscosity, measured in units of centistoke (cSt). The density of the material affects a direct conversion between centipoise and centistoke units, but for water and its density of 1, conversion is direct. You can compare the viscosity measurements of some common substances shown in Table 1.4 to a value listed in a resource and then have an understanding of the approximate viscosity of the particular material you are controlling.

More helpful in the field than measuring the viscosity of an unknown substance is the ability to alter the viscosity of a material in order to control its movement. Of secondary importance is the ability to confirm a pure substance by comparing its estimated viscosity to that listed in a reference source.

Viscosity can be lessened by mixing with a "thinner" liquid or heating. Viscosity can be increased by adding a solid, adding a polymer (starch), or cooling the material.

Water has a viscosity of 0.89 centipoise at 25°C. Liquids that are less viscous than water have an organic component.

Temperature of Product and Conversions

Temperature is simply a way to measure the heat energy within a substance. Atoms are in constant motion, vibrating and colliding with each other. Temperature is a way to measure that energy. The higher the collision rate, the higher the temperature (Figure 1.10).

There are three scales used to measure temperature: Fahrenheit, Celsius (centigrade), and Kelvin. Ongoing international attempts to precisely define Kelvin may make this explanation incomplete, but it's close enough for field use.

- Formula to convert from Fahrenheit to Celsius: (°F − 32) × 0.56 = °C.
- Formula to convert from Celsius to Fahrenheit: (°C × 1.8) + 32 = °F.
- Kelvin uses the same unit as Celsius, but is offset 273° because it uses absolute zero as a benchmark, not the freezing point of water.
- You can convert Celsius to Kelvin like this: °C + 273 = °K.

Chemical reactions are dependent on molecules colliding with each other. The faster the collision rate, the faster the rate of chemical reaction. But collision rate is not enough. The molecules colliding must have enough energy to start the reaction. As it turns out, the rate of chemical reaction will approximately double for each 10°C increase in temperature. This holds true in temperatures around room temperature. This is an approximation and is not as accurate at more extreme temperatures.

More specifically, the frequency of two particle collisions in gases is proportional to the square root of the temperature in Kelvin. For example, consider a ten-degree increase in temperature from room temperature (20°C). Convert the temperature to Kelvin, which is 20° plus 273, or 293°K. A 10°C increase is 303K. Divide 303 by 293 and then find the square root, which is 1.017. This ratio shows a 1.7% increase in the collision rate, but only a small

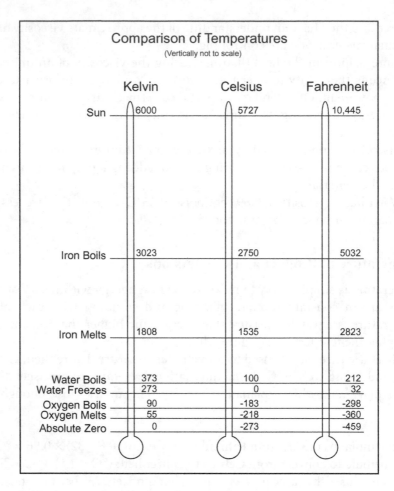

Figure 1.10 Comparison of temperature scales.

portion of those collisions contain enough activation energy to react. But it turns out that this small portion is roughly twice as many as those at the lower temperature. This is also why food stays fresh longer in a refrigerator, why an air monitor may not be accurate across wide ranges of temperature, and why colorimetric detector tubes and test strips will have a shorter shelf life if stored in a hot place.

Many other variables affect the rate of reaction, such as pressure, surface area, concentration, and catalyst.

Melting Point and Freezing Point

Melting point is the temperature at which a solid melts to a liquid. Freezing point is the temperature at which a liquid solidifies. Both terms indicate the same temperature for a material. Melting and freezing points assume standard pressure of 1 atmosphere, unless otherwise noted.

If a solid, such as wax, is heated in a metal bowl, the edges will melt first. As long as the heat is added slowly while stirring, the temperature of the melted wax will be the same as the block of solid wax in the center. The temperature of the liquid wax will not increase until all the solid wax is melted. Energy is required for this phase transition from a solid to a liquid. When the bowl contains only liquid wax, the temperature of the liquid will begin to climb.

The amount of energy required for the change from a solid to a liquid (or liquid to solid) is unique to each substance or mixture. For example, a block of ice will absorb heat at the rate of about 2 J/g/K. At the melting point, additional heat of 335 J/g is needed to complete the phase transition. Water will warm with the addition of about 4 J/g/K.

You can estimate the temperature of a material at its melting point if you observe both solid and liquid material intermixed. Most materials will present the solid phase as denser than the liquid phase due to contraction from the colder temperature. Most often, you will observe the solid phase submerged in the liquid.

Water presents a unique case when phase change is considered with density. Ice floats. Ice is unique in the way the molecules align to form a less dense shape. Warm water is less dense than cold water until about 37°F, when it reaches its maximum density. Water actually becomes less dense as it is cooled below 37°F and changes to ice.

You may encounter a hazardous material in a pond or other body of still water in cold conditions. If the water is more than a few feet deep with ice on top, know that the bottom of the pond is warmer than the water near the ice. Additionally, ground temperature across the United States seems to stabilize around 50–55°F several feet down. A pond in the winter has the influence of cold from above and heat from below. After the ice melts in the spring, the lowest, densest water at the bottom will absorb heat from the soil. When the lower layer reaches about 38–40°F, it will be less dense than the water above it and the pond will "turn over," in effect, agitating the entire body of water and sediment from the bottom.

Boiling Point and Condensation Point

Boiling point is the temperature at which a liquid turns to a gas. Condensation point is the temperature at which a gas condenses to a liquid. Both terms indicate the same temperature for a material. Boiling and condensation points assume standard pressure of 1 atmosphere unless otherwise noted.

The amount of energy required for the change from liquid to gas (or gas to liquid) is unique to each substance or mixture. For example, a pot of water will absorb heat at the rate of about 4 J/g/K. At the boiling point, additional heat of 2272 J/g is needed to complete the phase transition. Steam will warm with the addition of about 2 J/g/K.

The large amount of heat required for a phase transition from water to steam is what makes water an excellent quenching agent for fires and other hot objects. If you are applying water to a hot container and you observe the water boiling, you should know the temperature of the container is above 100°C, but the water is absorbing large amounts of heat.

Additionally, if an open container contains water that is boiling, you should know the temperature of the container in contact with the water is 212°F. This is analogous to a pot of water slowly boiling on a stove top. If the burner is turned up, the water boils more vigorously, but the temperature of the water will not increase.

Specific Heat

The specific heat of a substance is the amount of energy required to raise the temperature of one kilogram of that substance by one degree Celsius.

Thermal Energy

Thermal energy is composed of sensible heat and latent heart.

Sensible heat, or heat the can be sensed, is heat added or removed from a substance that changes the temperature.

Latent heat, or "hidden" heat, is the heat required for a substance to change phase, such as water to steam, and does not change the temperature. Latent heat of fusion is the heat required for a solid-to-liquid or liquid-to-solid phase change. Latent heat of vaporization is the heat required for a liquid-to-gas or gas-to-liquid phase change.

Generally, phase change requires more energy than raising the temperature of a substance.

Water makes a good example since everyone is familiar with ice, water, and steam. Figure 1.11 illustrates phase changes from ice through steam. The temperature does not rise until enough latent heat has been added to completely change the phase.

Heat Transfer

Heat transfer occurs through one of three ways:

- Conduction (direct contact)
- Convection
- Radiation

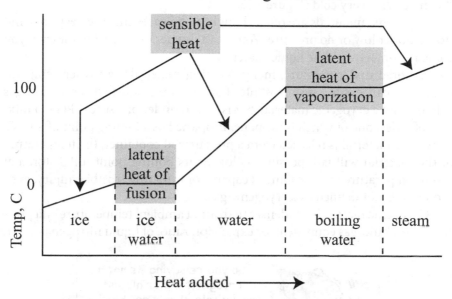

Figure 1.11 The relative relationship of phase change, sensible heat and latent heat for water. Specific heat value affects the slope of the lines depicting sensible heat.

Conduction is the transfer of heat between materials that are in contact. Some materials conduct heat readily, such as metal. Others conduct heat less readily (insulators), such as wood. The better the conductor, the faster heat can be transferred from one object to another.

Convection occurs within a substance, moving heat from hotter places to colder places. Warm portions of liquid or gas move within a container or the atmosphere to equalize temperature across the container. Convection causes currents within the container as long as heat is being transferred.

Conduction and convection both require matter to transfer heat.

Radiation does not require matter to transfer heat. Heat can be transmitted through empty space, including the space between air molecules, by thermal (infrared) radiation. The warmth of a campfire or the sun on a pleasant day is produced by infrared radiation.

Cryogenic

Cryogenic is an adjective that means "really cold." The term has various temperatures associated with definitions based on the needs of different groups working with particular substances. One definition of cryogenic gas is a material in liquid form with a boiling point of less than –150°F at 1 atmosphere of pressure. Other temperatures are used as a defining limit, but the point is, the material is very cold (Figure 1.12).

Cryogenic materials are often thought to be gases that are very cold and stored under low or no pressure. You are in danger of a freezing injury if you come in contact with the liquid material.

If stored under pressure, many escaping materials can reach temperatures that can easily cause frostbite. For example, liquefied petroleum gas (LPG) is not a cryogenic material, but is stored under pressure. LPG is a mixture of gases, one of which is propane. Propane has a boiling point of −42°C. When the material is released from a pressurized container, the temperature of the material will fall, possibly as low as its boiling point. LPG stored at room temperature and escaping is capable of inflicting frostbite injury, even though it is not defined as a cryogenic gas.

Cryogenic materials are denser than air at ambient temperature and pressure. The liquid/gas equivalent, or expansion ratio, of liquid nitrogen is 1:691,

Figure 1.12 Cryogenic material is defined as less than −150°F, which is extremely cold. Be aware of potential frostbite injury for any material colder than the freezing point of water.

so a gallon of liquid nitrogen at its boiling point will produce 691 gallons of nitrogen gas when adjusted to room temperature. Table 1.5 describes cryogenic gases by boiling point and expansion ratio.

Pressure and Units

The internationally recognized system for uniform description of units is the International System of Units (commonly abbreviated SI), from the French Systeme International d'unites. The SI unit for pressure is the Paschal (Pa). More commonly, pressure is measured in pounds per square inch (psi), atmospheres (atm), millimeters of mercury (mmHg), and other units. The pressure conversion values table may be used to compare pressures measured in differing units (Table 1.6). For example, 1 atmosphere (1 atm) is 14.696 pounds per square inch (psi) or 760 Torr (mmHg). The "=" symbol indicates the SI unit value in the table.

Standard Temperature and Pressure

Standard temperature, sometimes referred to as room temperature, is 68°F or 20°C and standard pressure is 1 atm (14.7 psi or 760 mmHg). STP is an abbreviation for "standard temperature and pressure."

Table 1.5   Cryogenic Gas by Increasing Boiling Point

Gas	Boiling Point,°C	Liquid/Gas Equivalent
Helium	−269.0	748
Hydrogen	−252.8	844
Neon	−246.1	1434
Nitrogen	−195.9	691
Carbon monoxide	−191.6	674
Fluorine	−188.1	946
Argon	−185.9	835
Oxygen	−183.0	854
Krypton	−153.4	699
Nitric oxide	−151.8	1040
Silane	−111.4	412
Xenon	−108.1	550
Ethylene	−103.8	482
Phosphine	−87.8	510
Hydrogen chloride	−85.1	772
Vinyl chloride	−13.8	365
Methylamine	−6.4	522
Phosgene	7.5	337

Source: Adapted from Air Liquid, SA website.

Table 1.6   Pressure Conversion Values

	Pound-Force per Square Inch (psi)	Atmosphere (atm)	Torr (mmHg)	Bar (bar)	Pascal (Pa)
1 psi	= 1 lb-f/in^2	0.068046	51.715	0.068948	6,894.76
1 atm	14.696	= 101 325 Pa	760	1.01325	101 325
1 Torr	0.019337	0.0013158	= 1 mmHg	0.0013332	133.322
1 bar	14.504	0.98692	750.06	= 100,000 dyn/cm^2	100 000
1 Pa	0.000145.04	0.0000098692	0.0075006	0.00001	= 1 N/m^2

Vapor Pressure

Vapor pressure is a measurement that you can use to predict a liquid's ability to drive vapor into the air. As a liquid pushes out vapor (e-*vapor*-ates), pressure is built up within its container and vapor pressure is an indicator of evaporation. A bloated plastic gas can during the summer is the result of approximately 3–5 psi (about 150–250 mmHg) vapor pressure from the liquid gasoline.

Vapor pressure is influenced by temperature. As the temperature rises, vapor pressure rises. When the vapor pressure equals atmospheric pressure, the liquid is at its boiling point.

The vapor pressure of water at standard temperature is about 18 mmHg. When the water is heated on a stove and boils, the vapor pressure equals atmospheric pressure, about 760 mmHg. The vapor pressure of acetone is 180 mmHg, so you can expect acetone to evaporate much more quickly than water under the same conditions.

Volatility

Volatility is the rate of evaporation of a liquid. Volatility is a relative term. The higher the vapor pressure, the more volatile a material becomes. A substance that is described as volatile at room temperature has a high vapor pressure, but there is no defining limit as to how high that pressure is. You may find it helpful to use water with its vapor pressure of 18 mmHg at room temperature as a benchmark when comparing volatile characteristics of other substances such as those found in Table 1.7.

Critical Temperature and Critical Pressure

You know that if you increase the temperature of a liquid, the vapor pressure increases and, in an open container, the evaporation rate increases. If a liquid is heated in a closed container, the vapor pressure increases and pressure builds. Conversely, if a gas is compressed enough in a container, it will

Table 1.7 Approximate Vapor Pressure of Common Substances

Substance	VP, mmHg @ 20°C
Hexanone	10
Octane	11
Water	18
Isopropanol	33
Ethanol	45
Butanone (MEK)	71
Methanol	98
Hexane	132
Acetone	184
Pentane	426
Ether	440

liquefy. If the container is opened, the liquid boils to a gas. Temperature and pressure interact to determine the phase of the material. However, above the critical temperature a different set of characteristics exist for any material.

Critical temperature is the temperature of a material above which distinct liquid and gas phases cannot be discerned. Critical pressure is the pressure produced at the critical temperature. At or above critical temperature, the properties of the gas and liquid phases become indistinguishable. In this phase, termed "supercritical fluid," the substance possesses characteristics of both gas and liquid. The substance can simultaneously diffuse through another material like a gas and dissolve other material like a liquid. This characteristic fulfills a niche in industrial applications.

During emergency response to a hazardous materials incident, a second characteristic of supercritical fluid becomes important. Above the critical temperature, a gas cannot be liquefied by pressure, which means explosive container failure may occur. The supercritical fluid can readily change density in response to small changes in temperature and pressure. Pressure may increase dramatically in response to continued heating. Supercritical fluid is less likely to absorb heat at the point of impingement on the container than would the corresponding liquid portion of a pressurized container.

Critical properties vary by material, just as is the case for the melting point, boiling point, and vapor pressure. Critical properties for many pure substances are listed in reference sources, but obtaining or estimating information for mixtures is more difficult to determine.

You can use critical temperature and pressure data as benchmarks when estimating conditions at a hazardous materials event (Table 1.8).

The freezing point of nitrogen at standard pressure is –346°F (–210.0°C, 63K). The boiling point of nitrogen at standard pressure is: –321°F (–195.8°C,

Table 1.8   Approximate Critical Temperature and Pressure Values

Substance	Critical Temperature, °C	Critical Pressure, atm
Acetylene	36	62
Ethylene oxide	196	71
Methanol	240	79
Ethanol	243	63
Isopropanol	235	47
Acetone	236	47
Butanone (MEK)	262	41
Chlorine	144	76
Ammonia	133	113
Oxygen	−118	50
Carbon dioxide	31	73
Nitrogen	−147	34

Source: Adapted from *CRC Handbook of Chemistry and Physics.*

77K). The critical temperature of nitrogen is −232°F (−47°C) and the critical pressure is 492.3 psi absolute (33.5 atm).

In other words, a block of frozen nitrogen, when warmed to −346°F, would melt and if warmed further to −321°F it would boil into a gas and exert 1 atm of pressure. If pressure was applied to this cold gas, it could be liquefied similar to liquefied petroleum gas in a cylinder. But if the pressurized cylinder of liquid was warmed to −232°F, the nitrogen would flash into a gas and no amount of additional pressure would be able to liquefy it. At the critical temperature of −232°F, the pressure in the container would be 33.5 atm (almost 500 psi).

Expansion Ratio

Expansion ratio describes the change in volume of a material as it moves from a liquid-to-gas phase without a change in temperature at standard pressure. The expansion ratio of liquefied petroleum gas (LPG) is about 1:270. In other words, 1 liter of LPG liquid expands to about 270 liters of vapor (Figure 1.13).

Expansion ratio is similar to "liquid/gas equivalent." Table 1.9 lists expansion ratio values for several industrial gases after stabilization at 15°C and 1 atm.

Many gases may be commercially liquefied by increased pressure, cooling, or a combination of both. Cryogenic fluids are characterized by having a boiling point of less than −150°F, but many other gases will liquefy above this temperature. You may see some materials labeled as "refrigerated" or "liquefied" and not necessarily "cryogenic."

Expansion Raito

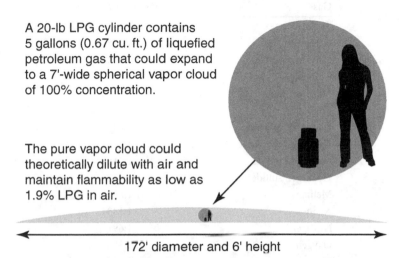

A 20-lb LPG cylinder contains 5 gallons (0.67 cu. ft.) of liquefied petroleum gas that could expand to a 7'-wide spherical vapor cloud of 100% concentration.

The pure vapor cloud could theoretically dilute with air and maintain flammability as low as 1.9% LPG in air.

172' diameter and 6' height

Figure 1.13 Scale illustration of the expansion of the liquefied contents of a full 20-pound LPG cylinder.

The boiling points of carbon dioxide and nitrogen are −78.5°C and −195.8°C, respectively. A physical property of cryogenic fluids and liquefied gases is the high volume expansion ratio in the liquid-to-gas phase. This ratio is 1.845 for carbon dioxide and 1:691 for nitrogen. These ratios were determined under very controlled conditions. In actuality, if a container of liquefied gas stored at room temperature were to break open and release the entire contents of the container, the results are not quite as clean as a simple expansion ratio suggests.

Imagine that a container at ambient temperature holding 1,000 gallons of anhydrous ammonia fractures and splits in half. How would the ammonia react? Table 1.10 provides data that can be used to predict the consequences of a catastrophic container failure. The liquid ammonia in the lower portion of the container would burst outward and begin to flash into ammonia gas. As gas formed, the remaining liquid would cool to its boiling point. Cold liquid would spill on the ground and nearby objects, gain heat from them, and flash more gas into the air. Heat energy from the air would flow into the colder ammonia and water vapor in the air would condense into a visible vapor cloud. Additionally, ammonia gas would dissolve into the water vapor in the air, forming caustic ammonium hydroxide mist. A plume would begin to move downwind and grow larger. Initially the plume would be cold and stay near the ground. As the plume warmed, it would begin to rise, move downwind, and dissipate at higher altitude. The plume would extend far beyond the visible cloud of condensed water vapor. As the water vapor

Table 1.9   Expansion Ratio of Various Gases

Gas	Expansion Ratio
Argon	1:835
Carbon dioxide	1:845
Carbon monoxide	1:764
Chlorine	1:521
Dichlorosilane	1:290
Ethylene	1:482
Hydrogen bromide	1:648
Hydrogen chloride	1:772
Hydrogen iodide	1:518
Hydrogen sulfide	1:638
Methane	1:630
Neon	1:1434
Nitrogen	1:691
Oxygen	1:854
Phosgene	1:337
Phosphine	1:510
Propane	1:311
Silane	1:412
Sulfur dioxide	1:535
Trifluoromethane (R23)	1:488
Vinylchloride	1:365
Xenon	1:550

Source: Adapted from AIR LIQUIDE, SA.

Table 1.10   Anhydrous Ammonia Properties

Ammonia, anhydrous Colorless gas with a pungent, suffocating odor [Note: Shipped as a liquefied compressed gas. Easily liquefied under pressure.]	DOT ID and Guide 1005/125 (anhydrous)
BP: –28°F FRZ: –108°F Sol: 34% VP: 8.5 atm RGasD*: 0.60	OSHA PELt: TWA 50 ppm (35 mg/m³) IDLH: 300 ppm Fl.P: NA (Gas) LEL: 15% UEL: 28%

Source: Adapted from NIOSH.

moved away from the release, it would warm and the condensation would also warm and become invisible again. The remaining ammonia liquid on the ground would cool the surrounding soil. The soil would slowly warm the liquid, which would be boiling. The liquid pool would continue to boil until all the liquid had evaporated or gone into solution with any water within the soil, in drains, or nearby surface water. Eventually, all the liquid would boil away, with the rate of boiling dependent on the rate of heat conducted by surrounding objects.

Auto Refrigeration

Auto refrigeration occurs when pressure is suddenly released from a container of pressurized or, more so, liquefied gas.

A pressurized gas will cool as it is allowed to expand from a pressurized container. This is observed when draining an SCBA or SCUBA cylinder; the container cools. Conversely, when the cylinder is filled with compressed air, the temperature rises. The increase or decrease in temperature returns to equilibrium with the environment at the new pressure.

A liquefied gas will exhibit more pronounced auto refrigeration than a pressurized gas. A reduction in pressure allows some of the liquefied material to change to gas. At that moment, the gas and liquid are still nearly the same temperature (the boiling point), but the act of changing from a liquid to gas form, a phase change, requires energy. The energy required for a phase change (freezing, melting, boiling or condensing) is usually much more than the energy required to raise or lower the temperature of the liquid by a degree. The energy for the phase change will often be expressed as a rapid cooling or heating of the material in the container (see latent heat, above).

Auto refrigeration of a liquefied gas is commonly observed with the discharge of a carbon dioxide fire extinguisher, which may frost the side of the container, or with the exuberant use of a propane grill, which will modestly cool the side of the propane tank. In either of these cases, the release of pressure is controlled.

The rapid or instantaneous release of pressure will produce a more severe cooling effect. An excellent recording of auto refrigeration during a 20 ton chlorine release was captured by the Jack Rabbit II project in the Utah desert at the U.S. Army Dugway Proving Grounds. The overall objective was to improve planning, tactical, operational and public protection actions during a catastrophic chlorine release or other TIH material.

Several excellent videos are available at https://www.uvu.edu/es/jack-rabbit/.

Sublimation

Sublimation is defined by a dictionary as:

> 1. (Chem.) The act or process of subliming, or the state or result of being sublimed; the process of vaporizing a solid and recondensing it into a solid, without it having first passed into the liquid state.

Certain solids, such as camphor, have a sufficiently high vapor pressure in the solid phase to make this a practical method for purification.

For hazmat response, sublimation illustrates the point that a compound may exist in more than one phase simultaneously. Common hazardous materials that can exist in more than one phase are iodine and mothballs.

Concentration

Concentration is a measure of how much of one substance is in another. It may apply to solid, liquid, or gas mixtures. Concentration is measured in several ways. It is usually expressed in terms of volume or weight. A solution becomes less concentrated if more solvent is added and the solution becomes more concentrated if more solute is added.

As mentioned previously, solution concentration may be expressed as a percentage based on the ratio of mass per mass, mass per volume, or volume per volume. Percentages are based on a scale of 1/100th. Other units may be used for greater precision. Beyond percent (parts per hundred), are ppm (parts per million or 1/1,000,000) and ppb (parts per billion or 1/1,000,000,000). One percent equals 10,000 ppm or 10,000,000 ppb.

When the particles are not uniformly shaped or distributed evenly, a unit of milligrams per cubic meter (mg/m^3) is used. For the purpose of calculating or predicting, unevenly distributed particles measured in milligrams per cubic meter may be converted to parts per million by a simple calculation. The formula is from the National Institute for Occupational Safety and Health (NIOSH) and allows you to calculate one of three values if the other two values are known (Table 1.11). The formula assumes ambient conditions of 25°C and 1 atmosphere.

Concentration may also be expressed by mole (mol). This allows you to refer to the number of particles in a solution regardless of the size or mass of the particle.

One mole of a substance contains Avogadro's number of particles, or about 6.022×10^{23}. One mole of a substance is that quantity of a substance when mass in grams is the same as its atomic weight. For example, to determine one mole of sodium hydroxide (NaOH), add the molecular weights of sodium, oxygen, and hydrogen. The total atomic weight is:

$$22.99 + 16.00 + 1.01 = 40.00$$

Table 1.11   NIOSH Conversion Calculator

Theoretically convert concentrations measured in parts per million to or from milligrams per cubic meter with the following formula:
X ppm = (Y mg/m^3)(24.45)/(molecular weight)
or Y mg/m^3 = (X ppm)(molecular weight)/24.45

Source: Adapted from NIOSH.

This means one mole of sodium hydroxide weighs 40 grams and consists of 22.99 grams of sodium (1 mol), 16.00 grams of oxygen (1 mol), and 1.01 grams of hydrogen (1 mol). There are about 6.022×10^{23} each of sodium atoms, oxygen atoms, and hydrogen atoms combining to form those 40 grams. Additionally, if 40 grams of sodium hydroxide is dissolved in water, it dissociates into 1 mol of sodium cation and 1 mol of hydroxide anion.

Molarity (M) is a unit of concentration that is expressed as moles per liter of solution. Forty grams of sodium hydroxide (1 mol) in 1 liter of water is a 1 M (one molar) solution. Halve the number of particles or double the volume of solvent to get a 0.5 M solution. The traditional designation of M is being replaced by the approved SI unit of mol/L or mol/m^3.

Molarity expresses concentration in moles per volume. Molality (m) expresses concentration in terms of moles per mass, or mol/kg. One mole of particles in 1 kilogram of solvent is a 1 m (molal) solution. The traditional designation of m is being replaced by the approved SI unit of mol/kg or mol/g.

Molarity is calculated using the volume of the entire solution, but molality is calculated using the mass of solvent only.

"Normality" is a term referring to the phenomenon of salts that dissociate in solution. These dissociated ions exist in solution but cannot be isolated and then measured. One normal (N) is one gram equivalent of solute per liter of solution. One gram equivalent is based on the number of protons or electrons that may be transferred. For example, a 1 mol/L solution of sodium hydroxide [NaOH] is a 1 N solution based on its ability to provide one hydroxide anion per molecule. But a 1 mol/L solution of calcium hydroxide [Ca(OH)$_2$] is a 2 N solution based on its ability to provide two hydroxide anions per molecule.

Normality is used to measure a single ion in a solution, such as sodium or hydroxide, but not the solution itself. You will find normality used most often to describe the strength of acids and bases; however, normality also describes other salts.

Normality is helpful in acid base neutralization reactions because it considers the proportion of strength to volume. When normality is applied to a known volume of solution, the corresponding volume of neutralizing solution may be calculated:

$$N_a V_a = N_b V_b$$

Vapor Density

Vapor density is similar to specific gravity, only it applies to vapors and gases, not solids or liquids. Vapor density is a ratio of a solid or liquid to air. Air is used as the standard and is given a value of 1. Something with a vapor density greater than 1 will sink in air. Less than 1 will float. For example, acetone has a vapor density of 2.0, so acetone vapor will sink because it is two times heavier than air. Ammonia has a vapor density of 0.6, so ammonia vapors will rise. Some resources use other terms for vapor density, such as specific gravity (gas). The NIOSH Pocket Guide to Chemical Hazards uses RGasD, relative gas density.

Temperature will affect these values, so they are only accurate if the vapor and the air are the same temperature. A vapor that is hotter than air will initially rise and will then tend to sink or rise, depending on the vapor density once the vapor is the same temperature as the air (see Figure 1.14).

Physical Change

Physical changes are basically any changes to a material that are not chemical changes. If a material changes phase (solid, liquid and gas), changes volume, produces light (infrared and fluorescence), shatters, floats, sinks, absorbs another material, rolls down a hill, etc., it is a physical change. Compare to "Chemical Change and Reactivity."

Vapor Density, Temperature and Plumes

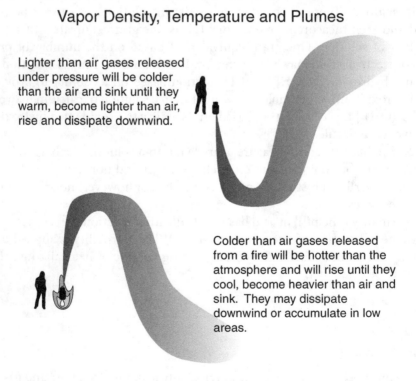

Lighter than air gases released under pressure will be colder than the air and sink until they warm, become lighter than air, rise and dissipate downwind.

Colder than air gases released from a fire will be hotter than the atmosphere and will rise until they cool, become heavier than air and sink. They may dissipate downwind or accumulate in low areas.

Figure 1.14 Illustration of a heated heavier-than-air chlorine plume dissipating downwind and a cold lighter-than-air anhydrous ammonia plume dissipating downwind.

Flash Point

Flash point is the lowest temperature of a flammable liquid that can form an ignitable mixture with air. According to the National Fire Protection Association (NFPA), flash point is "the minimum temperature at which a liquid or a solid emits vapor sufficient to form an ignitable mixture with air near the surface of the liquid or the solid." NIOSH defines flash point as "the temperature at which the liquid phase gives off enough vapor to flash when exposed to an external ignition source." The vapor pressure of the flammable liquid determines the output of flammable vapor. The vapor pressure is determined by the liquid temperature.

The concentration of flammable vapors in air must be within a range in order for a flame to propagate through the mixture of vapor and air. As each fuel molecule reacts with oxygen in the air, the heat produced by the reaction is passed to more fuel and oxygen. If the fuel molecules are too far apart, as in low concentration mixtures, too little energy is passed to the next fuel-oxygen pair to incite ignition. In order for a flash to occur, the liquid must be heated to a temperature at which vapor production is able to produce a concentration in air that is high enough to propagate the flash. The flash will extinguish when the fuel-air mixture is consumed.

Flammable liquids require different fuel-air concentrations to flash and extinguish, so each flammable liquid as well as each mixture will have differing flash points. Flash point applies to solids, but they tend to be much higher. "Flash point" is a term that is usually applied to flammable and combustible liquids.

The U.S. Department of Transportation (DOT) defines flammable and combustible liquids by class, which includes parameters of temperature, phase, etc. There are other definitions for flammable and combustible liquids and solids based on use, storage, and shipping conditions. Unless associated with a definition, "flammable" and "combustible" simply mean the substance will burn. Inflammable means the same thing as flammable – that is, the material is capable of burning. Inflammable does not mean the substance is unflammable. The word "inflammable" is fading from use due to the confusion the prefix causes.

Flash point is determined in a laboratory using a device that uniformly heats the flammable liquid. Several devices may be used: Cleveland, Pensky-Martens, Setaflash, and others (Figure 1.15). An ignition source is placed just above the liquid as the liquid temperature slowly increases. Flash point is recorded as the temperature at which vapors first ignite or flash. The device may use a closed cup or an open cup to hold the liquid as it is heated. Closed cup may be abbreviated as c.c. and open cup as o.c. Closed-cup values are often slightly lower than open cup values because the open-cup method allows more ventilation and dissipation of flammable vapors.

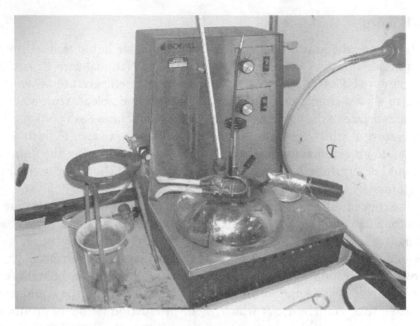

Figure 1.15 This device is used for determination of flash point using the Pensky-Martens closed-cup method.

Table 1.12   Classification of Flammable and Combustible Liquids

Class	Flash Point Range and Boiling Point Range
IA flammable liquid	FP < 73°F and BP < 100°F
IB flammable liquid	FP < 73°F and BP at or > 100°F
IC flammable liquid	FP > 73°F and < 100°F
II combustible liquid	FP > 100°F and < 140°F
IIIA combustible liquid	FP > 140°F and < 200°F
IIIB combustible liquid	FP > 200°F

Source: Adapted from OSHA criteria.

Flash point is used to classify fire hazards of flammable and combustible liquids (Table 1.12). Flash points of common substances are listed in Table 1.13. Flash point is not related to the temperature of the ignition source or the temperature of the air.

Fire Point

Fire point is the lowest temperature at which a liquid open to the atmosphere will produce sufficient vapors to continue to burn when exposed to an ignition source. Usually, the fire point is slightly higher than the flash point for the same liquid.

Table 1.13   Flash Point of
Common Substances*

Substance	Flash Point (°F)
Diesel fuel	100–130
Hexane	–10
Gasoline	–45
Kerosene	100–160
Methanol	52
Corn oil	489
Propane	–156
Acetone	1
Toluene	40

*Common substances are generic
mixtures. Values are approximate.

At the fire point, the temperature of the liquid produces more than enough vapors to flash. The higher concentration of burning vapor in the air produces additional heat energy, which heats the liquid. The additional heating produces vapors at a rate fast enough to continue to feed the fire. The flash becomes a self-sustaining fire that will not extinguish due to lack of fuel in the air as a flash fire would.

Ignition (Auto Ignition) Temperature

Ignition temperature is the temperature of a substance at which it will ignite in air without the presence of an ignition source. Ignition temperature is usually much higher than the flash point or fire point (Table 1.14). Table 1.15 summarizes the definitions for the flash point, fire point, and ignition point.

Table 1.14   Ignition Point and Flash Point
of Common Substances*

Substance	Ignition Point (°F)	Flash Point (°F)
Diesel fuel	177–263	100–130
Hexane	453	–10
Gasoline	495	–45
Kerosene	536	100–160
Methanol	725	52
Corn oil	738	489
Propane	842	–156°F
Acetone	869	–4
Toluene	896	40

*Common substances are generic mixtures. Some
values are approximate.

Table 1.15   Comparison of Definitions

Term	Temperature at Which a Material ...
Flash point	... flashes, but does not sustain fire, with an ignition source.
Fire point	... ignites and continues to burn when exposed to an ignition source.
Ignition (auto-ignition) point	... ignites and continues to burn without exposure to an ignition source.

Flammable (Explosive) Range (LEL and UEL)

LEL (lower explosive limit) is the lowest concentration of flammable vapor in air that will burn. UEL (upper explosive limit) is the highest concentration of flammable vapor in air that will burn. The flammable range spans from LEL to UEL.

Combustible vapors in air may be produced by solid, liquid, or gas material. Without a fire, the source material will produce vapors with the maximum concentration of vapors occurring at the source of the release. In the case of a high vapor pressure flammable liquid, vapors are very concentrated at the surface of the liquid. Lower vapor pressure liquids are likely to have less concentrated vapors at the surface. In the case of a release of flammable liquid without fire, liquid will flow to a low point and produce vapors. Vapor production is a function of temperature. Soon equilibrium is reached and vapor concentration is highest at the surface of the liquid and lessens as the vapors move downwind and dissipate. Somewhere between the liquid surface and the far reaches of the atmosphere, the concentration of flammable vapors and oxygen from ambient air exists in the right proportions to allow a fire to begin if an ignition source is present. The only time flammable vapors would not accumulate in sufficient concentration just above the surface of the liquid is if the liquid is below its flash point. A typical plume is illustrated in Figure 1.16.

The LEL is defined as the lowest concentration of a flammable vapor that can burn in air. At low concentrations, there is too much space between flammable molecules to pass heat energy onto the next set of molecules and a fire cannot propagate.

The UEL is defined as the highest concentration of flammable vapor in air that can burn. High concentrations of flammable vapor will dilute the concentration of oxygen in ambient air, which has a suffocating effect on the reaction and a fire cannot propagate.

LEL and UEL are measured in percent concentration in air. The flammable range, sometimes called explosive range, spans LEL to UEL. Resources may list LEL as a percent measurement, UEL as a percent measurement, or flammable range as a percent range. For example, if flammable range is listed as 2.4–4.5, you should recognize the LEL as 2.4% and the UEL as 4.5%.

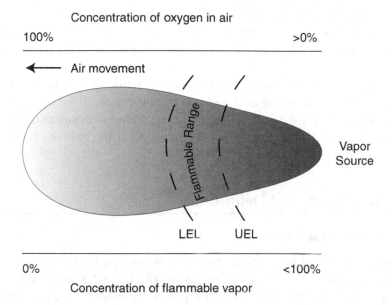

Figure 1.16 Flammable range illustration.

Table 1.16 Flammable Range (percent) of Common Substances*

Diesel fuel	0.6–7.5
Hexane	1.1–7.5
Gasoline	1.4–7.6
Kerosene	0.6–4.9
Methanol	6.0–36.0
Methane	5.0–15.0
Propane	3.0–13.0
Acetone	2.1–9.5
Toluene	1.2–7.1

*Common substances may be generic mixtures. Some values are approximate.

Flammable range can be determined for individual substances as well as mixtures (Table 1.16).

Contamination vs. Exposure

Contamination means you are still in contact with the contaminant and you need to be decontaminated. Exposure means you were exposed to the harmful effects of a hazard for a period of time, but are no longer exposed because

you left the area. For example, if you walk into an area where a "dirty" radiological bomb has been detonated, you may be exposed to radiation. If you
leave the area, you are no longer exposed. However, if you contaminate your
hands with radioactive dust, you have been contaminated. You can leave the
area, but you are still being exposed to radiation as the dust clings to your
hands.

In the same way, breathing gasoline vapors while fueling your car is an
exposure. You can leave the area and end the exposure. If you spill gasoline
on your hand or clothing, you will take some gasoline with you and continue
your exposure until you decontaminate by washing the gasoline from your
skin or removing the contaminated clothing.

Chemical Terms and Definitions

Chemical Interaction

Chemical interaction occurs when two or more compounds come in contact
and produce a chemical change. The change can be instantaneous, gradual,
spontaneous, or initiated by a third compound or energy. Chemical interactions produce lower net energy states. Sometimes additional energy supplied
by fire, electricity, light, or other energy source must be added to initiate a
chemical reaction.

Chemical Change and Reactivity

Chemical change describes chemical interactions or reactions. Chemical
change involves a rearrangement of atoms within molecules along with new
bonding schemes to form new materials. Energy is consumed or produced by
the change and the change may occur slowly or instantaneously. Chemical
reactions either require or release energy. If the energy is heat, endothermic
means the reaction requires heat and exothermic means the reaction liber-
ates heat.

Highly reactive compounds tend to release large amounts of energy
quickly. Reactivity is affected by concentration, temperature, pressure, surface area, the presence of contaminants, and other physical characteristics.
For example, a block of magnesium submerged in water is merely wet. The
same block of magnesium ground into a fine powder and wetted with water
will begin to react and self-heat. The heat will increase the rate of the reaction
to the point of ignition.

Most sources use reactivity in a relative manner. The Environmental
Protection Agency (EPA) defines a specific set of conditions to define reactivity. Reactive waste is designated D003 in the EPA classification system used
to characterize waste, but not product.

According to the EPA, a solid waste exhibits the characteristic of reactivity if a representative sample of the waste has any of the following properties:

- It is normally unstable and readily undergoes violent change without detonating.
- It reacts violently with water.
- It forms potentially explosive mixtures with water.
- When mixed with water, it generates toxic gases, vapors, or fumes in a quantity sufficient to present a danger to human health or the environment.
- It is a cyanide- or sulfide-bearing waste that, when exposed to pH conditions between 2 and 12.5, can generate toxic gases, vapors, or fumes in a quantity sufficient to present a danger to human health or the environment.
- It is capable of detonation or explosive reaction if it is subjected to a strong initiating source or if heated under confinement.
- It is readily capable of detonation or explosive decomposition or reaction at standard temperature and pressure.
- It is a forbidden explosive as defined in 49 CFR 173.51, or a Class A explosive as defined in 49 CFR 173.53, or a Class B explosive as defined in 49 CFR 173.88.

The Canadian Centre for Occupational Health and Safety uses Workplace Hazardous Material Information System (WHMIS) criteria to define dangerously reactive liquids and solids as those that can undergo vigorous polymerization, condensation, or decomposition; become self-reactive under conditions of shock or increase in pressure or temperature; and react vigorously with water to release a lethal gas.

Instability

Instability is a relative measure of a compound's tendency to react. "Instability" is a term that is often used to describe the ability to control the compound, not to grade the power of a reaction. For example, the instability of homemade nitroglycerine is greater than most military grade explosives, even though military grade explosives tend to be much more powerful.

Remember that uncontrolled chemical reactions will be driven by the reactants reconfiguring to a lower energy state with the excess energy usually being released as heat. Instability describes the relative ease with which the degrading reaction occurs and does not indicate how quickly it occurs. A commercial product may be described as instable because it has a short shelf life, not because it is highly reactive.

Instability of a substance may be illustrated by a sled perched at the top of a snow covered hill. Some amount of energy must be expended to begin the reaction and thus reach a lower energy state. The longer and steeper the hill, the more energy is released, but instability does not describe the long run down the hill, just the amount of work it takes to begin the slide. Figure 1.17 shows that A is more instable than B because it takes less added energy to initiate the reaction.

Very instable compounds can react vigorously or explosively under conditions of impact such as a hammer blow or even slightly elevated temperature or pressure. Some of the very instable substances are:

- Ammonium perchlorate
- Azo and diazo compounds
- Acetylides
- Azides
- Fulminates
- Hydrogen peroxide solutions (91% by weight)
- Many organic peroxides
- Nitro and nitroso compounds
- Nitrate esters
- Perchloric acid solutions (over 72.5% by weight)
- Picric acid
- Picrate salts
- Triazines

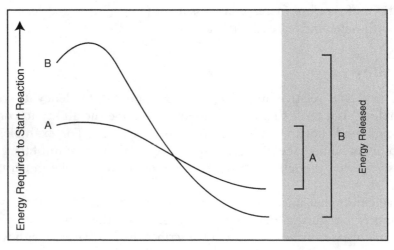

Relative Instability of Compounds

Figure 1.17 A is more instable than B because it takes less added energy to initiate the reaction.

Ionic and Covalent Compounds

Ionic and covalent compounds are named by the way atoms bond to each other within the molecule.

Ionic compounds bond ionically, that is, there is a positively charged ion (a cation) and a negatively charged ion (an anion). They stick together like two magnets. Metals from the left side of the periodic table form cations and elements from the right side of the periodic table form anions. Sometimes several atoms form a complex ion with a net positive or negative charge. Since it is the entire ion that carries the charge, several negative atoms can overcome the positive charge of a single metal cation. This allows some metals to exist as part of a complex anion such as chromate (CrO_4^{-2}) or permanganate (MnO_4^-).

Ionically bonded compounds, often referred to as salts, are usually solids with high boiling points because these strong bonds form a lattice or crystal line structure.

Ionic compound nomenclature describes the cation first and ends in "-ium." The anion is described second. If it is a simple anion, which contains only one element, the second word usually ends in "-ide." If the anion is complex, that is, it contains more than one element, it may end in "-ide" or it may end in "-ous," "-ite," or "-ate," depending on the oxidation state of the anion.

Covalent compounds bond by sharing electrons rather than transferring electrons like ionic compounds do. Covalent bonds occur mainly between nonmetals found on the right side of the periodic table. By far, covalent bonds are found in organic (hydrocarbon) chemistry.

Organic and Inorganic

Organic chemistry involves structures of carbon and hydrogen covalently bonded to form hydrocarbons. By definition, any compound that contains covalently bonded hydrogen and carbon is a hydrocarbon. Organic chemistry is involved in the chemistry of life. Our bodies are mainly organic if water is ignored.

Inorganic compounds include all that is not organic. Inorganic compounds make up rocks, water, and air. Inorganic compounds may be ionically or covalently bonded. Some of these covalently bonded compounds were named long ago with a different naming system and cause a little confusion now and again. Carbon dioxide, for example, is covalently bonded even though it ends in "-ide" and it is not a hydrocarbon because it contains only carbon and not hydrogen.

Generally speaking, organic compounds are lighter than water and can burn. Inorganic compounds often look like rocks or dirt and are usually heavier than water.

Saturated, Unsaturated, and Aromatic Hydrocarbons

Carbon forms the structure of a hydrocarbon and links to other carbons in chains. Each carbon forms four bonds, either to other carbons, to hydrogen, or to some functional group.

If a chain of carbons is linked together with just one bond between each carbon, the remaining bonding sites are filled with hydrogen. In other words, the carbon chain is saturated with hydrogen. If someone refers to a saturated hydrocarbon, you can assume the carbons are all linked with single bonds.

Straight and branched chains refer to the carbon to carbon structure of the molecule drawn as an illustration. Hydrogen is often ignored in these illustrations. Straight chains look like a telephone pole and branched chains look like a tree.

Organic covalent bonds contain energy for reactions. Single carbon-carbon bonds are the most stable. If the carbons are linked somewhere with a double carbon bond, it means there are not as many places for hydrogen to fit. The presence of a carbon = carbon double bond means the carbon chain is no longer saturated with hydrogen, so it is termed an unsaturated hydrocarbon. If someone refers to an unsaturated hydrocarbon, you can assume the molecule contains at least one carbon = carbon bond, which is more reactive than a carbon-carbon bond. Triple bonds are likely to produce a more violent reaction than a double bond.

Halogenated Hydrocarbon

A halogenated hydrocarbon is any hydrocarbon that contains a covalently bonded halogen. Halogens are found in column 7 of the periodic table and include, commonly, fluorine, chlorine, bromine, iodine, and, uncommonly, astatine.

Halogenated hydrocarbons tend to be insoluble in water and heavier than water. They also tend to resist burning and extinguish fire. Hydrocarbons are lighter than water and are able to burn. Halogenated hydrocarbons also tend to destroy ozone in the upper atmosphere. The characteristic of each halogenated hydrocarbon depends on the size of the hydrocarbon portion of the molecule as well as the size and number of the halogen.

Halogenated hydrocarbons are recognized by name by the presence of one or more halogen prefixes coupled to a hydrocarbon name. More specifically, they may be classed according to the halogen. For example, fluoromethane is a hydrofluorocarbon. Carbon tetrachloride is a chlorocarbon. 1,2 bromoethane is a hydrobromocarbon.

Halogenated hydrocarbons are useful in several applications. They have strong solvent properties, such as methylene chloride. Perchloroethylene is dry cleaning fluid. Halon® is a trade name for a set of halogenated hydrocarbons with strong fire suppression characteristics. Another set of halogenated hydrocarbons are used as refrigerants and are known by the trade name Freon®. Several other products are listed in Table 1.17.

Table 1.17   Common Names of Halogenated Hydrocarbons

Chemical Name	Common Name	ANSI/ASHRAE Code
1,1,1,2,3,3,3-Heptafluoropropane		HFC-227ea
1,1,1,2-Tetrafluoroethane	R-134a	HFC-134
1,1,1-Trichloro-2,2,2-trifluoroethane		CFC-113a
1,1,1-Trichloroethane	Methyl chloroform	
1,1,2,2,2-pentafluoroethane	Pentafluoroethane	HFC-125
1,1,2,2-Tetrafluoro ethane		HFC-134
1,1,2-Trichloro-1,2,2-trifluoroethane	Trichlorotrifluoroethane	CFC-113
1,1-Dichloro-1-fluoroethane		HCFC-141b
1,1-Dichloroethane	Ethylidene dichloride	Freon 150a
1,1-Difluoroethane		HFC-152a
1,2-Dichlorotetrafluoroethane	Dichlorotetrafluoroethane	CFC-114
1,2-Dichloroethane	Ethylene dichloride	Freon 150
1-Chloro-1,1,2,2,2-pentafluoroethane	Chloropentafluoroethane	CFC-115
1-Chloro-1,1-difluoroethane		HCFC-142b
2-Chloro-1,1,1,2-tetrafluoroethane		HFC-124
Bromochlorodifluoromethane	Halon 1211	Halon 1211
Bromotrifluoromethane	Halon 1301	Halon 1301
Chlorodifluoromethane	R-22	HCFC-22
Chloromethane	Methyl chloride	
Chlorotrifluoromethane		CFC-13
Decafluorobutane	perfluorobutane	R610
Dibromomethane	Methylene bromide	
Dichlorodifluoromethane	Freon-12, R-12	CFC-12
Dichloromethane	Methylene chloride	
Difluoromethane		HFC-32
Fluoromethane	Methyl fluoride	HFC-41
Hexachloroethane		CFC-110
Polybrominated biphenyl	PBB	
Polychlorinated biphenyl	PCB	
Polychloroethene	polyvinyl chloride, PVC	
Polytetrafluoroethene	PTFE, Teflon	
Tetrachloromethane	Carbon tetrachloride	CFC-10
Tetrafluoromethane	Carbon tetrafluoride	CFC-14
Tribromomethane	Bromoform	
Trichlorofluoromethane	Freon-11, R-11	CFC-11
Trichloromethane	Chloroform	
Trifluoroio domethane	Trifluoromethyl iodide	Freon 13T1
Trifluoromethane	Fluoroform	HFC-23

Source: Adapted from Environmental Protection Agency website.

Polymerization

Polymerization is the chemical reaction that forms plastics and other large webs of molecules when controlled. When uncontrolled, many polymerization reactions will produce enough heat to burn or explode.

Monomer means "one part" and polymer means "many parts." Plastics and other polymers are formed when monomers combine.

Monomers react with each other to form a huge web of polymer or maybe only a chain of a few units in length. Much of the data available on polymerization involve properties of reactions used to make useful products. The type of polymerization you should be concerned with is hazardous polymerization.

Hazardous polymerization reactions from monomers produce heat, pressure in enclosed containers, and toxic vapors. These types of monomers are reactive, instable, and chemically interactive alone or with other chemicals. Polymerization reactions may release enough energy to ignite.

Monomers may react due to a variety of conditions. Some monomers will auto polymerize, that is, react with themselves. Others will react with other monomers. Some will form peroxides, which build to a concentration that can initiate hazardous polymerization. Others will initiate a reaction in the presence of initiators, catalysts, or contaminants.

Some polymerization reactions are harmless, such as gravy formed from starch. Polymers are very common in biological systems since proteins (polymers) are formed from amino acids (monomers) and starches are formed from sugars. Hazardous monomer and polymer chemicals are often recognized by "poly" in the name, such as polystyrene from styrene monomer, polyvinyl chloride from vinyl chloride monomer, and others. Some monomers are inhibited or stabilized for transport and may include these words in the name, for example, "Styrene, Inhibited." Some monomers will use oxygen from the air to form peroxides, which exacerbates the polymerization hazard. Other monomers are packaged with a small amount of inhibitor, which must have oxygen available to prevent polymerization.

Phenol-formaldehyde reactions are commonly used in the production of plastics. The reaction of phenol or a substitute with an aldehyde in the presence of a catalyst is used to prepare phenolic resins. Phenolic resins are used in adhesives, coatings, and other compounds. The type of catalyst used, the ratio of reagents, and heat and pressure determine the properties of the resin. These reactions produce large amounts of heat and are sensitive to physical and chemical conditions. Once the reaction begins, heat generated by the reaction increases the rate of reaction, generating more heat in a continuous feedback. The rate of heat generation will accelerate because the reaction rate is an exponential function of the temperature. A runaway reaction can occur if the large amount of heat produced cannot be dissipated. As the reaction begins to accelerate, the pressure of the system will increase suddenly due to gas production and the induced evaporation of liquid. Excess pressure can exceed the capacity of the container and cause an explosion.

A case study involves plastic producer Georgia-Pacific in 1997 when an 8,000 gallon batch reactor exploded while producing phenolic resin. A report from the EPA reports the following:

> An operator charged raw materials and catalyst to the reactor and turned on steam to heat the contents. A high temperature alarm sounded and the operator turned off the steam. Shortly after, there was a large, highly energetic explosion that separated the top of the reactor from the shell. The top landed 400 feet away. The shell of the reactor split and unrolled, and impacted against other vessels. A nearby holding tank was destroyed and another reactor was partially damaged. The explosion killed the operator and left four other workers injured.
>
> The investigation revealed that the reactor explosion was caused by excessive pressure generated by a runaway reaction. The runaway was triggered when, contrary to standard operating procedures, all the raw materials and catalyst were charged to the reactor at once followed by the addition of heat. Under the runaway conditions, heat generated exceeded the cooling capacity of the system and the pressure generated could not be vented through the emergency relief system causing the reactor to explode.

Polymerization reactions are not as uncommon as you might expect. Table 1.18 lists a short history of just phenol-formaldehyde runaway polymerization reactions. Then Table 1.19 presents a list of monomers and corresponding stabilizers.

Self-Accelerating Decomposition Temperature (SADT)

All organic peroxides slowly decompose because they are inherently unstable. The rate of decomposition is dependent on the temperature and is unique to each organic peroxide. A characteristic of the decomposition reaction is the release of heat, which hastens further decomposition. Self-accelerating decomposition temperature (SADT) is the temperature at which the organic peroxide will produce heat faster than it can dissipate. This is defined as $\geq 6°C$ in one week. Most organic peroxide SADTs range from −10 to 200°C. Smaller containers have a larger surface area to the mass within and are therefore more sensitive to heating because they cannot dissipate the heat as well as a larger mass.

As the temperature rises, the decomposition accelerates and feeds more heat into the product. The resulting accelerating decomposition temperature can result in an intense fire, a fireball, or an explosion. Organic peroxides can produce an explosive equivalent of about 5–40% that of TNT.

Table 1.18   Phenol-Formaldehyde Reaction Incidents at Various Companies

Date of Incident	Description	Effects
September 10, 1997	An 8,000 gallon reactor exploded during production of a phenol-formaldehyde resin.	1 worker fatality, 4 employees injured, 3 firefighters treated for chemical burns. Evacuation of residents for several hours.
August 18, 1994	Pressure buildup during manufacture of phenolic resin, pressure increased, rupture disks popped. Product was released through emergency vent. The cause of accident was reported as failure to open condensate return line.	Residents evacuated for 5 hours.
February 29, 1992	A 13,000 gallon reactor exploded during production of a phenol-formaldehyde resin. Explosion occurred during initial stages of catalyst addition.	4 employees injured, 1 seriously. 1 firefighter treated for chemical burns. Evacuation of 200 residents for 3 hours.
November 11, 1991	Temperature increased in chemical reactor, releasing phenol-formaldehyde resin.	None reported.
October 16, 1989	Manufacture of phenolic resins and thermoset plastics; release of phenol and formaldehyde from process vessel.	None reported.
August 28, 1989	Manufacture of phenolic resins; release of phenol and phenolic resin from process vessel; "operator error" cited as cause.	1 injured.
July 25, 1989	Specialty paper manufacturing; release of phenolic resin and methanol from process vessel.	None reported.

Table 1.19   Partial List of Monomers and Stabilizers

Monomer	Stabilizer	Concentration
Acrylic acid	MEHQ	200 ppm
Acrylonitrile	Monomethyl ether hydroquinone	35–45 ppm
Butadiene	p-tert-butylcatechol	Various
Chlorotrifluoroethylene	Tributylamine	1%
Ethylene	NA	NA
Methyl methacrylate	Monomethyl ether hydroquinone	10–100 ppm
Styrene	4-tert-butylcatechol	10–15 ppm
Tetrafluoroethylene	Various terpenes	Variable
Vinyl acetate	Hydroquinone	3–20 ppm
Vinyl chloride	Phenol	–

Source: Table adapted from various safety data sheets and resources.

Many organic peroxides require some type of refrigeration or temperature control. A manufacturer may list in the safety data sheet (SDS) a recommended storage temperature that will optimize the shelf life of the product. It may also list a minimum storage temperature that will prevent crystallization or phase separation that would increase reactivity. The maximum storage temperature specifies a safe storage temperature, but extended storage at this temperature will degrade the peroxide over several weeks or months without developing an SADT. The emergency temperature is set at 20°C (36°F) below the lowest SADTs and may be a lesser value for higher SADTs. Planned, preventative action is initiated at the emergency temperature and evacuation occurs at the SADT.

Most organic peroxides will act as fuels (reducers). As organic peroxides decompose, most do not release free oxygen from the oxygen-oxygen bond (–O–O–) and often carbon dioxide is released. However, organic peroxides do release free radicals, which accelerate the decomposition and encourage auto ignition of the hydrocarbon portion of the molecule as it is sprayed into the air. The peroxy acids are an exception to this and are oxidizing.

Some organic peroxides are inhibited with an additive. Other products are mixtures of organic peroxides and other compounds. Organic peroxides can often be recognized by "peroxide" at the end of the name or "peroxy" in the name.

Maximum Safe Storage Temperature (MSST)

The maximum safe storage temperature is the highest temperature at which a product may be stored before hazardous decomposition begins.

Catalyst

A catalyst is a substance used to speed up a chemical reaction that would otherwise run slowly or not at all. A catalyst is not consumed by the reaction and remains after the reaction, although it may or may not be an intermediary in the reaction. Sometimes a catalyst can initiate a reaction while other cases require an initiator to start the reaction and a catalyst to continue it. Catalysts may be heterogeneous, homogenous, or biological.

A heterogeneous catalyst provides a surface for the reaction to take place. The catalyst is finely divided metal powder or some other platform. The catalyst positions the reactant(s) in a manner that facilitates the reaction. The most effective catalysts tend to be transition metals. A vehicle catalytic converter is a heterogeneous catalyst that contains platinum or sometimes rhodium and aids in the conversion of carbon monoxide and nitric oxide toxins to less toxic carbon dioxide and nitrogen.

A homogenous catalyst is a molecule that facilitates the reaction but is then released or reformed. The catalyst may or may not be changed during the reaction, but the same amount of catalyst remains after the reaction.

A biological catalyst is an enzyme. Countless enzymes facilitate biochemical reactions that make life possible. Ricin toxin is an enzyme that is able to ratchet its way down nerve pathways and destroy nerve protein via catalysis at the rate of 50,000 units per second.

While a catalyst is used to initiate a reaction, inhibitors are used to stabilize a reactive chemical or slow the rate of a chemical reaction.

Inhibitor

An inhibitor is a substance that prevents or decreases the rate of a chemical reaction. It may operate by tempering the rate of reaction between two or more reagents. Inhibitors can also slow or prevent reactions by blocking the reaction site on a catalyst or enzyme.

Inhibitors in materials may slowly decrease during storage even when stored at recommended temperatures. At storage temperatures higher than recommended, inhibitor levels can decrease at a much faster rate. At storage temperatures lower than recommended, the inhibitors may separate with subsequent loss of inhibition function. Some inhibitors need oxygen to work effectively.

Vapors evaporating from inhibited material do not contain the inhibitor. If these vapors condense, form polymers, and expand, they can clog a vent or pressure relief device.

Fire Triangle

The fire triangle has three sides: fuel, air, and heat (Figure 1.18). A more advanced model of the fire triangle is:

- Fuel (reducer)
- Air (oxidizer)
- Heat (energy)

Fuel may be eliminated by shutting off the fuel source to the fire, for example, closing a gas valve. The fire may be denied air, such as when a burning pan of oil is covered with a lid. Heat may be eliminated from a fire by the quenching ability of water or some other cooling agent. In practice, fire fighting often employs more than one method.

The fire tetrahedron model includes a fourth aspect, the chemical chain reaction that is the propagation of the flame. Halon and ammonium

The Fire Triangle

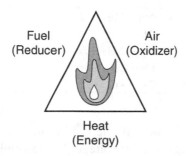

Fuel
(Reducer)

Air
(Oxidizer)

Heat
(Energy)

Figure 1.18 The fire triangle.

phosphate work in part to extinguish fire by inhibiting the chemical chain reaction. As an inhibitor, these agents interfere with the flow of energy within the fire, which reduces the efficiency of the flame. The point here is that fire is inhibited in the presence of these compounds.

Oxidation Potential, Oxidizers, Oxyanions, and Organic Peroxides

Oxidizers can be oxygen, provide oxygen, or act like oxygen in a fire. Oxidizers are not combustible but they enhance the rate of combustion. Some oxyanions, such as nitrate, perchlorate, and peroxide, can also be oxidizers. You would be right to think that in order to have a powerful fire, you would need to use an extremely flammable fuel in combination with air. Another way to produce a powerful fire is to use a common fuel and increase the power of the oxidizer. The stronger the oxidizer is in combination with a fuel, the more spectacular the combustion.

A redox reaction is a term used to describe electron flow in a reaction. Redox is from reduction-oxidation reaction. An oxidizer (like oxygen) gives electrons and a reducer (like fuel) takes electrons. Redox reactions can describe many chemical reactions from slow rusting to flaming explosions.

Oxidation potential is a way to measure the energy transferred between the oxidizer and reducer in a redox reaction. Energy released during these reactions may or may not be sufficient to support a fire. Measurements of oxidation potential are relative to each other, so they are benchmarked to a specified substance. In Table 1.20, the benchmark is the energy required to ionize hydrogen and the oxidation potential is measured in volts (V). The benchmark is arbitrarily designated as zero and the others are measured by the voltage needed to induce the change from neutral to ionized based on the energy needed to induce the change in hydrogen.

Table 1.20   Oxidation Potential

	Compound	Oxidation Potential (V)
Stronger reducers	$Mg \rightarrow Mg_{+++} 2e^-$	2.37
	$Al \rightarrow Al^{+++} + 3e^-$	1.66
Benchmark	$H_2 \rightarrow 2 H^+ + 2e^-$	0.00
	$Cu - Cu^{++} + 2e^-$	−0.34
Stronger oxidizers	$2 Cl^- - Cl_2 + 2e^-$	−1.36

The greater the difference in the oxidation potential value, the more energy is released as measured by voltage. For example, if copper is used to oxidize aluminum, the difference in the oxidation potential values is 1.66 − (−0.34) = 2.00. If chlorine is used to oxidize aluminum, the difference is 1.66 − (−1.36) = 3.02. Chlorine is the more powerful oxidizing substance because the difference in oxidation potential is greater.

References list oxidation and reduction potential for many compounds based on several benchmarks and they are useful for chemists and industry in the application of pure substances. However, these are not very useful values when characterizing hazards of an unknown substance, especially if the substance in question is a mixture or contaminated with other material. What is more useful is for you to understand and apply the concept that oxidizers and reducers are relative to each other and it takes a wide range in oxidation potential to produce a fire.

Oxidation ability is a relative term describing how actively a compound may participate as an oxidizer. Reduction ability is the inverse of oxidation ability.

More concentrated oxidizers accelerate combustion. Thirty-five percent hydrogen peroxide can cause spontaneous combustion and is self-concentrating. (VP H_2O_2 = 5 mmHg. VP H_2O = 18 mmHg.) A list of oxidizing substances as defined by the U.S. Department of Transportation is shown in Table 1.21.

Organic peroxides are special oxidizers in that they contain an oxidizing portion and a reducing portion in the same molecule. In other words, an organic peroxide molecule contains both fuel and oxidizer tied together. The missing side of the fire triangle is the energy portion necessary to initiate combustion. When peroxides degrade, it is often carbon dioxide that is released, not oxygen. The decomposition of the organic peroxide may be violent enough to burn or explode. For the fire to spread to other nearby objects, atmospheric oxygen is necessary to support combustion.

DOT classification of organic peroxide is based on the substance's type (A–F), phase (solid or liquid), and whether it requires temperature control (Table 1.22).

Table 1.21 DOT Oxidizers, Class 5.1

Aluminum nitrate

Ammonium dichromate

Ammonium nitrate based fertilizer

Ammonium nitrate emulsion or suspension

Ammonium nitrate, liquid [hot concentrated solution]

Ammonium nitrate, with < 0.2% total combustible material …

Ammonium perchlorate

Ammonium persulfate

Barium bromate

Barium chlorate, solid

Barium chlorate, solution

Barium hypochlorite [with more than 22% available chlorine]

Barium nitrate

Barium perchlorate, solid

Barium perchlorate, solution

Barium permanganate

Barium peroxide

Beryllium nitrate

Bromates, inorganic, aqueous solution, n.o.s.

Bromates, inorganic, n.o.s.

Bromine pentafluoride

Bromine trifluoride

Calcium chlorate

Calcium chlorate aqueous solution

Calcium chlorite

Calcium hypochlorite, dry [or] calcium hypochlorite mixtures dry …

Calcium hypochlorite, hydrated [or] mixtures

Calcium hypochlorite mixtures, dry, with more than 10% …

Calcium nitrate Calcium perchlorate

Calcium perchlorate

Calcium permanganate

Calcium peroxide

Cesium nitrate [or]

Caesium nitrate

Chlorate and borate mixtures

Chlorate and magnesium chloride mixture, solid

Chlorate and magnesium chloride mixture solution

Chlorate and magnesium chloride mixture, solid

Chlorates, inorganic, aqueous solution, n.o.s.

Chlorates, inorganic, n.o.s.

Chloric acid aqueous solution, [with not more than 10% chloric acid]

Chlorine dioxide, hydrate, frozen

Chlorites, inorganic, n.o.s.

(Continued)

Table 1.21 (Continued)   DOT Oxidizers, Class 5.1

Chromium nitrate

Chromium trioxide, anhydrous

Copper chlorate

Dichloroisocyanuric acid, dry [or] Dichloroisocyanuric acid salts

Didymium nitrate

Ferric nitrate

Guanidine nitrate

Hydrogen peroxide and peroxyacetic acid mixtures, stabilized.

Hydrogen peroxide, aqueous solutions (with more than 40% but not …)

Hydrogen peroxide, aqueous solutions (with not less than 20% but not …)

Hydrogen peroxide, aqueous solutions (with not less than 8% but less …)

Hydrogen peroxide, stabilized [or] aqueous solutions …

Hypochlorites, inorganic, n.o.s.

Iodine pentafluoride

Lead dioxide

Lead nitrate

Lead perchlorate, solid

Lead perchlorate, solution

Lithium hypochlorite, dry [with > 39% available chlorine (8.8% …)]

Lithium nitrate

Lithium peroxide

Magnesium bromate

Magnesium chlorate

Magnesium nitrate

Magnesium perchlorate

Magnesium peroxide

Manganese nitrate

Nickel nitrate

Nitrates, inorganic, aqueous solution, n.o.s.

Nitrates, inorganic, n.o.s.

Nitrites, inorganic, aqueous solution, n.o.s.

Nitrites, inorganic, n.o.s.

Oxidizing liquid, corrosive, n.o.s.

Oxidizing liquid, n.o.s.

Oxidizing liquid, toxic, n.o.s.

Oxidizing solid, corrosive, n.o.s.

Oxidizing solid, flammable, n.o.s.

Oxidizing solid, n.o.s.

Oxidizing solid, self-heating, n.o.s.

Oxidizing solid, toxic, n.o.s.

Oxidizing solid, water reactive, n.o.s.

Oxygen generator, chemical [including when contained …]

Perchlorates, inorganic, aqueous solution, n.o.s.

(Continued)

Table 1.21 (Continued) DOT Oxidizers, Class 5.1

Perchlorates, inorganic, n.o.s.

Perchloric acid [with more than 50% but not more than 72% …]

Permanganates, inorganic, aqueous solution, n.o.s.

Permanganates, inorganic, n.o.s.

Peroxides, inorganic, n.o.s.

Persulfates, inorganic, aqueous solution, n.o.s.

Persulfates, inorganic, n.o.s.

Potassium bromate

Potassium chlorate

Potassium chlorate, aqueous solution

Potassium nitrate

Potassium nitrate and sodium nitrite mixtures

Potassium nitrite

Potassium perchlorate

Potassium permanganate

Potassium peroxide

Potassium persulfate

Potassium superoxide

Silver nitrate

Sodium bromate

Sodium carbonate peroxyhydrate

Sodium chlorate

Sodium chlorate, aqueous solution

Sodium chlorite

Sodium nitrate

Sodium nitrate and potassium nitrate mixtures

Sodium nitrite

Sodium perborate monohydrate

Sodium perchlorate

Sodium permanganate

Sodium peroxide

Sodium peroxoborate, anhydrous

Sodium persulfate

Sodium superoxide

Strontium chlorate

Strontium nitrate

Strontium perchlorate

Strontium peroxide

Tetranitromethane

Thallium chlorate

Trichloroisocyanuric acid, dry

Urea hydrogen peroxide

Zinc ammonium nitrite

(Continued)

Table 1.21 (Continued)   DOT Oxidizers, Class 5.1
_____

Zinc bromate

Zinc chlorate

Zinc nitrate

Zinc permanganate

Zinc peroxide

Zirconium nitrate
_____

Source: Adapted from U.S. Department of Transportation.

Fuel/Oxidizer Combinations and the Fire Triangle

"Normal" fuels can cause spectacular reactions when combined with a high concentration of strong oxidizers. Common hydrocarbons, such as asphalt or oil, can become explosive when saturated with liquid oxygen. Alcohol, kerosene, and other common hydrocarbon fuels combined with hydrogen peroxide (H_2O_2) or nitrous oxide (N_2O) will produce intense combustion as concentration increases.

Rocket engines use powerful oxidizers in combination with powerful fuels. One combination is liquid hydrogen and liquid oxygen. Another is nitrogen tetroxide and hydrazine. Some oxidizer/fuel combinations will spontaneously ignite or explode.

A 1997 case study by the EPA documented a fatality from mixing waste solvents and fuels with oxidizers to heat a cement production kiln. On the day of the incident, two workers were on top of the disperser, a large tank with a rotating paddle used to mix waste. They were pouring liquids from 55 gallon drums into the disperser. They were starting a new batch and only four drums of liquid had been added to the tank when a lab employee at the top of the tank added one bucket of chlorates, one bucket of perchlorates, and one bucket of nitrites to the disperser. Each five-gallon bucket contained about three or four inches of dry material. The mixer was not running at this time. Thirty to 60 seconds after the oxidizers were added, and while waste from a fifth drum was being dumped into the tank, liquid suddenly erupted back out of the large tank opening, followed by an explosion and fireball. The fireball fatally engulfed the employee who was pouring the drums and started a large fire in the building. The fire spread to other flammable materials stored throughout the building.

Key findings from the incident indicate the mixture of fuels and oxidizers as well as the manner in which they were mixed were responsible for the reaction. Although the exact chemical mechanism is not precisely known, given the chemicals present in the disperser residue, a violent reaction could have occurred through several possible mechanisms. The immediate cause of the explosion and fire was most likely a violent reaction of oxidizers in the disperser in the presence of flammable liquid and vapor. Since only four drums had been dumped into the previously empty disperser, only about nine inches of liquid would have been in the bottom of the tank, or about half of the amount needed to reach the mixer. This allowed the solid oxidizers to pile up at the bottom of the tank in direct contact with each other and with flammable solvent liquid and vapor. The high concentration of oxidizer moistened by solvent could not dissipate heat as readily as compared to mixing while the disperser was operating (Figure 1.19).

Table 1.22 Classifications of Organic Peroxides

Type	Description
A	Can detonate or deflagrate rapidly as packaged for transport. Transportation is forbidden.
B	When packaged for transport, neither detonates nor deflagrates rapidly, but can undergo a thermal explosion.
C	When packaged for transport, neither detonates nor deflagrates rapidly and cannot undergo a thermal explosion.
D	Detonates only partially, but does not deflagrate rapidly and is not affected by heat when confined; does not detonate, deflagrates slowly, and shows no violent effect if heated when confined, or does not detonate or deflagrate, and shows a medium effect when heated under confinement.
E	Neither detonates nor deflagrates and shows low, or no, effect when heated under confinement.
F	Will not detonate in a cavitated state, does not deflagrate, shows only a low, or no, effect if heated when confined, and has low, or no, explosive power.
G	Will not detonate in a cavitated state, will not deflagrate at all, shows no effect when heated under confinement, and shows no explosive power.

Source: Adapted from U.S. Department of Transportation.

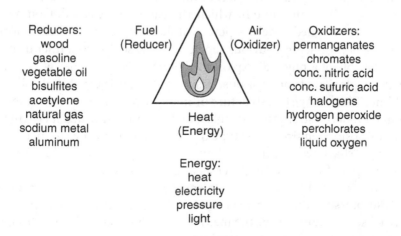

Reducers:
wood
gasoline
vegetable oil
bisulfites
acetylene
natural gas
sodium metal
aluminum

Fuel
(Reducer)

Air
(Oxidizer)

Oxidizers:
permanganates
chromates
conc. nitric acid
conc. sufuric acid
halogens
hydrogen peroxide
perchlorates
liquid oxygen

Heat
(Energy)

Energy:
heat
electricity
pressure
light

Figure 1.19 Examples of reducers, oxidizers, and energy related to the fire triangle.

Toxic Products of Combustion

All products of combustion are toxic. The poison is in the dose.

Smoke contains individual toxic gases such as hydrogen cyanide, carbon monoxide, oxides of nitrogen, etc., as well as particulates. Generally, less toxic products of combustion are produced by high temperature fire as opposed to a cooler, partial combustion.

Explosives

Explosives are designed to produce a shock wave and sudden expansion of gas. A reducer and oxidizer are integrated in a single compound. Explosives don't necessarily burn; many decompose to a gas. For example, many nitrogen-containing compounds convert from a solid to a gas by shifting electrons and liberating nitrogen or other gases. The speed of the reaction is nearly instantaneous and can cause a shock wave traveling up to 35,000 feet per second.

Explosives perform work primarily through the shock wave. A shock wave from an explosion consists of a wave of energy moving first in a compression wave and then in an expansion wave. The shock wave produces rapid pressure changes in any material. The material, or medium, affects the speed of the wave and the way energy is absorbed and transmitted through the medium. Fragmentation and heat are also hazards.

Explosive decomposition that produces a shock wave that travels less than the speed of sound in air (about 761 miles per hour or 340 meters per second) is a deflagration. A detonation occurs when the shock wave moves faster than sound. A detonation is a more powerful explosion because the shock wave moves through the explosive material, compressing the material as the explosive reacts.

An explosion may cause damage or injury based on the amount of energy it releases as well as the time in which the energy is released. Commercial explosives are designed to deliver powerful shock waves, but other explosions are capable of producing injurious shock waves. A sealed container heated in a fire is capable of a physical explosion that will dissipate energy as a shock wave. An explosion of vapor or dust suspended in air is capable of collapsing a building. The damage to a structure or injury to a person is determined by a combination of the strength of the blast and the time in which it is delivered.

An explosion requires some type of containment in order to increase the brisance or "sharpness" of the shock wave bursting out in a sudden manner. Consider a pot of water boiling over a campfire and a can of soup in the fire. The pot of water will slowly dissipate energy while boiling. The can of soup will build pressure until it bursts. Materials designed to be explosive may only need surrounding air or the mass of explosive materials itself to initiate and propagate an explosive shock wave.

Table 1.23 lists the detonation velocities at specified (typically, the highest practical) density of various explosive compounds. The velocity of detonation is an important indicator for overall energy or power of detonation, and in particular for the brisance or shattering effect of an explosive.

Air Reactivity

Air-reactive substances react when exposed to air. These materials react violently in contact with air or oxygen or with compounds containing oxygen. However, air contains other chemicals, one of which is water vapor. Nitrogen and the trace gases are largely inert. Air-reactive substances may produce significant heat, fire, or toxic gases. End products of the reaction are likely

Table 1.23   Explosive Detonation Velocities

Explosive	Abbreviation	Detonation Velocity (m/s)
Silver azide		4,000
Mercury fulminate		4,250
Lead azide		4,630
Lead styphnate		5,200
Ammonium Nitrate	AN	5,270
Methyl nitrate		6,300
Ethyl picrate		6,500
Methyl picrate		6,800
Trinitrocresol		6,850
Trinitrotoluene	TNT	6,900
Ammonium picrate (ammonium picrate)	Explosive D	7,150
Picryl chloride		7,200
Trinitrobenzene	TNB	7,300
Trinitroaniline	TNA	7,300
Nitroglycol (ethylene glycol dinitrate)	EGDN	7,300
Picric acid		7,350
Triaminotrinitrobenzene	TATB	7,350
Tetryl (2,4,6 tetranitro-N-methylaniline)		7,570
Ethylenedinitramine	EDNA	7,570
Nitroglycerine	Nitroglycerine	7,600
Nitroguanidine		8,200
Mannitol hexanitrate		8,260
Pentaerythritol tetranitrate	PETN	8,400
Cyclotrimethylenetrinitramine	RDX	8,750
Cyclotetramethylene tetranitramine	HMX	9,100
Tetranitroglycoluril	Sorguyl	9,150

Source: Adapted from Explosives Engineering.

to be hazardous. Air-reactive substances are pyrophoric, self-heating, or self-reacting.

Pyrophoric materials will burn when exposed to air. Self-heating materials will react when exposed to air without an external energy source. These materials are also referred to as spontaneously combustible. The flame of some pyrophoric materials is clear and not readily visible.

The U.S. Department of Energy (DOE) defines "pyrophoric materials" as those which will immediately burst into flame when exposed to air. DOE defines "spontaneously combustible materials" as those that will ignite after a slow buildup of heat.

DOT defines "pyrophoric material" as a solid or liquid material that will burn without an ignition source within five minutes of exposure to air according to United Nations specifications. DOT defines self-heating materials as those that exceed 200°C or self-ignite within 24 hours under certain test criteria. These specifications are for defining purposes as they relate to the safe transportation of dangerous goods and are not intended for characterizing hazards of unlabeled substances.

Corrosive (Acids and Bases/Alkaline) and Dissociation

Corrosive substances are those capable of slowly destroying material, usually through chemical action of excess hydronium or hydroxide ions. A DOT corrosive is a substance that has a high corrosion rate on steel or human tissue. It is a definition concerned with transportation issues and actually contains both acids and bases.

Generally, an acid is a substance that produces excess H^+ in water and a base is a substance that produces excess OH^- in water. One H^+ and one OH^- combine to form H_2O in a neutralization reaction. Acidic and basic solutions, as measured by pH, occur when the balance of H^+ and OH^- is altered from the natural balance of ions in water, which is one ion per 10,000,000 water molecules, or pH 7. The equilibrium is described as:

$$2H_2O \rightleftharpoons H_3O^+ + OH^-$$

It is significant to note that equilibrium describes the entire solution and the equation above is heavily weighted to the left. Every time this dissociation happens, it does not happen to ten million other water molecules.

Acids are also called "corrosives" and described as being "acidic." Acids often have the term "acid" at the end of the name. Bases are also called caustics and are described as being "alkaline" (from this term they are sometimes called an "alkali"). They are sometimes recognized by the name "hydroxide" at the end of the name.

Acids were found long ago to taste sour and produce a stinging sensation. Bases tasted bitter and produce a slippery sensation as they change tissue fat to soap in the skin. As stronger and more concentrated acids and bases were developed, these hazardous characteristics increased and injury could be severe and immediate. Acids and bases will permeate skin. The most effective treatment for acid or base exposure is 15–20 minutes of flushing with clean water in order to draw the acid or base from the skin.

The DOT groups acids and bases as "Class 8, Corrosive." The DOT definition simply defines corrosives in transportation as substances that can corrode human tissue or metal at a rate of 0.25 inch per year. The term "corrosive" as used by DOT includes both corrosives and caustics lumped into a single class.

There are three basic definitions of an acid according to Arrhenius, Brønsted-Lowry, and Lewis. Arrhenius is the simplest and most useful for hazardous characterization. Lewis is the most complex and most useful in research and industry.

According to the Arrhenius definition, an acid is a substance that increases the concentration of hydronium ion (H_3O^+ or more commonly, H^+) when dissolved in water, while bases are substances that increase the concentration of hydroxide ions (OH^-). This definition limits acids and bases to substances that can dissolve in water. This definition requires measurement in a wet or water environment.

According to the Brønsted-Lowry definition, an acid is a proton (H^+) donor and a base is a proton acceptor. The acid is said to be dissociated after the proton is donated. Brønsted and Lowry formulated this definition to include water insoluble substances and solutions that do not contain water.

According to the Lewis definition, an acid is an electron pair acceptor and a base is an electron pair donor. Lewis acids are electrophiles and Lewis bases are nucleophiles. The Lewis definition is used extensively in organic chemistry because a molecule that is an electrophile (electron lover) can have an area of partial positive charge, which will attract an electron. A nucleophile has a partial negative charge, which attracts an electrophile. Lewis acids include substances with no transferable H^+ and a Lewis base includes substances with no transferable OH^-.

Presumably, you are identifying hazardous characteristics of an unknown substance in order to protect, in decreasing priority, people, property, and the environment. Since people are mostly water and water is abundant in our environment, it makes sense to consider acid and base characteristics in contact with water containing substances, like people, while being aware of the other definitions.

A common wastewater treatment chemical, ferric chloride, is a Lewis acid and has no available H^+ or OH^-, yet it is labeled as a DOT Class 8 Corrosive with a statement that reads "Danger! Corrosive. Causes burns to any area of contact." While pure iron chloride will not produce an accurate reading on a pH strip, the addition of a little water will produce an accurate reading. By identifying hazardous characteristics in the context of a water-based system, you can easily identify corrosive properties.

Some acids can provide more than one H^+ per molecule of acid. Sulfuric acid (H_2SO_4) is an example. Some bases can likewise produce more than one (OH^-) per molecule. This fact becomes important when predicting neutralization reactions.

It is important to note that pH only measures the concentration of H^+ in solution that has already dissociated. An acid that falls apart in water to release nearly all its H^+ into water is a strong acid because it can dissociate easily. But the term "strong" acid does not tell you how concentrated the H^+ is. pH measures concentration and reveals how strong the solution is. For example, hydrochloric acid is a strong acid and acetic acid is a weak acid. Adding a drop of concentrated hydrochloric acid into a swimming pool will produce a solution of strong acid with weak strength. Making a 3% solution of acetic acid in water (similar to vinegar) will produce a solution of weak acid with more corrosive strength than the swimming pool solution.

Inorganic acids tend to dissociate more completely than organic acids, so inorganic acids tend to be stronger than organic acids. Because of this, acids can be grouped into strong, moderate, and weak groups (Table 1.24). It is important to note that regardless of the term "weak" or "strong," the strength of an individual solution can only be measured by pH.

Bases are the counterpart to acids. You can think of them as "opposites" to keep things simple. Bases, like acids, are described as strong or weak based on the dissociation ability of the base (Table 1.25). A solution is basic if it has a higher than normal (pH 7) concentration of hydroxide ion (OH^-). This is the same as having a lower than normal hydronium (H^+) ion concentration.

Table 1.24   Relative Strength of Some Common Acids

Strong Inorganic Acids	Moderate to Weak Inorganic Acids	Weak Organic Acids
Hydrobromic acid	Boric acid	Acetic acid
Hydrochloric acid	Carbonic acid	Benzoic acid
Hydroiodic acid	Chloric acid	Butyric acid
Nitric acid	Hydrofluoric acid	Citric acid
Sulfuric acid	Phosphoric acid	Formic acid
Perchloric acid	Pyrophosphoric acid	Lactic acid

Table 1.25   Relative Strength of Some Common Bases

Strong Inorganic Bases	Moderate to Weak Inorganic Bases	Weak Organic Bases
Potassium hydroxide	Ammonium hydroxide	Methylamine
Barium hydroxide	Sodium carbonate	Pyridine
Sodium hydroxide	Sodium bicarbonate	
Calcium hydroxide	Magnesium hydroxide	
Lithium hydroxide		

Both sodium carbonate and ammonia are bases, although neither of these substances contains OH^- groups. That is because both compounds accept H^+ when dissolved in water.

pH

pH is a measurement of the concentration of hydronium ion (H^+) in water. Water is H_2O and has a neutral pH of 7. Occasionally, one water molecule will dissociate into H^+ and OH^-. This natural dissociation produces a concentration of one H^+ for every 10,000,000 water molecules. There is also one OH^- for every 10,000,000 water molecules.

The pH scale is a way to measure the imbalance of H^+ or OH^- in a water solution. The concentration of the ions is measured in M, or equivalent moles per liter. The natural dissociation of H^+ produces a concentration that is measured as the inverse logarithm of the H^+ concentration.

Ten million is 10^7 and pH is measured on an inverse logarithmic scale. The "7" in pH neutral water is indicated by the inverse log. You can see that a pH 7 solution is "neutral," but still contains H^+ and OH^-. Acidic solutions have a pH less than 7 and basic solutions have a pH greater than 7.

Solutions do not become strongly acidic or basic unless the pH is toward the ends of the scale. Solutions develop strongly acidic or basic pH values when the natural dissociation of water becomes imbalanced. pH values for some common materials are shown in Figure 1.20.

Strength

"Strength" refers to the amount of corrosive work a solution can do. Since each pH unit is a "power of ten," the difference in strength of a one pH unit move from 1 to 0 is much greater than a move from 5 to 4.

Neutralization

"Neutralization" means, literally, to make something ineffective – to counteract or nullify. It is most often applied, in this setting, to acid/base reactions,

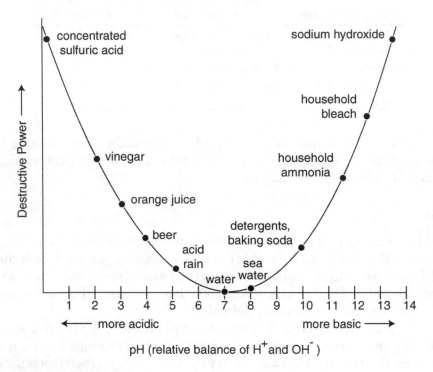

pH and Destructive Power of Common Substances

Figure 1.20 pH values for some common materials.

but may be used in other situations, such as reduction/oxidation reactions or electrical/magnetic phenomenon.

When an acid and base combine, the net effect is a combination of H⁺ and OH⁻ to form water, a salt that may or may not stay in solution, and heat.

Neutralization of Acids

When dissolved in water, the base sodium hydroxide decomposes into hydroxide and sodium ions:

$$NaOH \rightarrow Na^+ + OH^-$$

Likewise, hydrogen chloride in water forms hydronium and chloride ions:

$$HCl + H_2O \rightarrow H_3O^+ + Cl^-$$

When the two solutions are mixed, the H₃O+ and OH⁻ ions combine to form water:

$$H_3O^+ + OH^- \rightarrow 2H_2O$$

If equal quantities (moles) of NaOH and HCl are dissolved, the base and the acid neutralize each other and only sodium chloride (NaCl), a salt, remains in the solution.

Equal amounts (moles) of acid and base are needed for neutralization reactions between strong acids and strong bases. This formula describes the moles needed in proportion to the concentration of the acid and base.

$$a \times [A] \times V_a = b \times [B] \times V_b$$

The number of hydrogen ions is a, while b is the constant that tells you how many H_3O^+ ions the base can accept. [A] is the concentration of acid and [B] is the concentration of base. V_a is the volume of acid and V_b is the volume of base.

WMD Terms and Definitions

Biological Agents and Biological Toxins

Biological agents and toxins are life forms or substances obtained from life forms that have been manipulated in a way to make them more toxic or lethal.

Manipulations may include collecting an especially potent strain of bacteria or virus, concentrating or purifying an agent or toxin, freeze drying, coating, and others.

Biological agents and toxins include:

- Bacteria (anthrax, cholera, diphtheria, plague), which are single-celled organisms used to cause disease in humans and animals by invading tissue or producing toxins. A bacterium contains a nucleus with DNA, protein, and other biochemicals unique to the organism or characteristic of a group of organisms. Antibiotics are pharmaceuticals used to kill bacteria.
- Viruses (Ebola, smallpox, Dengue and Rift Valley fever) are microorganisms that depend on a host to replicate. Viruses have a protein coat that protects the viral genetic material. The virus must invade and control a host in order to hijack its replicating mechanism. Viruses are not sensitive to antibiotics. Some pharmaceuticals will slow the rate of viral reproduction but not kill the virus.
- Rickettsiae (typhus, Rocky Mountain spotted fever) are similar to bacteria but have more viral properties. Rickettsiae must invade a host cell to replicate. They are sensitive to antibiotics.

- Fungi (coccidioidomycosis, mushrooms, molds, mildews, smuts, rusts, yeasts) are a diverse group of organisms that lack chlorophyll and vascular tissue and may be a single cell or a mass of branching filaments called hyphae that produce a fruit. A limited number of anitfungal pharmaceuticals exist.
- Toxins (anatoxins, ricin, saxitoxin, botulinum) are chemicals produced by organisms. Toxins do not reproduce and are not sensitive to antibiotics. Many toxins achieve high toxicity through enzymatic action within a human, animal, or plant.

Table 1.26 lists biological agents and toxins that are deemed threats to humans and agricultural activities. The U.S. Department of Health and Human Services (HHS) column lists those biological agents and toxins that pose a direct threat to humans under the heading "HHS Select Agents and Toxins." The United States Department of Agriculture (USDA) lists those biological agents and toxins that pose a direct threat to animals and plants important to agricultural activity. Animal threats are listed under the heading "USDA select agents and toxins" and plant threats are listed under "USDA plant protection and quarantine (PPQ) select agents and toxins." Agents and toxins that pose a threat to both areas are listed as "overlap select agents and toxins."

Biological agents and toxins kill or incapacitate over time by overwhelming infection, enzyme action, or some other biological function.

Dose

Dose means literally "the act of giving." So if a person receives a dose, it would be measured as the total amount of something received in a situation.

Dose Response

Dose response is the exposed subject's expression of a toxic effect. This is different than the dose rate, the rate at which the dose is given (not expressed). The dose is usually proportional to the effect in the person exposed, but the dose rate must be considered, too. For example, inhaling one anthrax spore is not likely to be fatal. Inhaling 5,000 spores is fatal to an average person and about twice that dose if the person is very fit. But what if an average person inhaled ten spores per day for 500 days for a total of 5,000 spores? Wool sorters have been observed to inhale about 75 spores per hour with no significant effect. The combination of dose and time may be used to describe a dose response curve, which is seldom linear.

Table 1.26 Select Biological Agents and Toxins

HHS Select Agents and Toxins	Overlap Select Agents and Toxins	USDA Select Agents and Toxins
Abrin	*Bacillus anthracis*	African horse sickness virus
Cercopithecine herpesvirus 1 (Herpes B virus)	Botulinum neurotoxins	African swine fever virus
Coccidioides posadasii	Botulinum neurotoxin-producing species of *Clostridium*	Akabane virus
Conotoxins	*Brucella abortus*	Avian influenza virus (highly pathogenic)
Crimean-Congo hemorrhagic fever virus	*Brucella melitensis*	Bluetongue virus (exotic)
Diacetoxyscirpenol	*Brucella suis*	Bovine spongiform encephalopathy agent
Ebola virus	*Burkholderia mallei* (formerly *Pseudomonas mallei*)	Camel pox virus
Lassa fever virus	*Burkholderia pseudomallei* (formerly *Pseudomonas pseudomallei*)	Classical swine fever virus
Marburg virus	*Clostridium perfringens* epsilon toxin	*Cowdria ruminantium* (Heartwater)
Monkeypox virus	*Coccidioides immitis*	Foot and mouth disease virus
Reconstructed replication competent forms of the 1918 pandemic influenza virus containing any portion of the coding regions of all eight gene segments (reconstructed 1918 influenza virus)	*Coxiella burnetii*	Goat pox virus
Ricin	Eastern equine encephalitis virus	Japanese encephalitis virus
Rickettsia prowazekii	*Francisella tularensis*	Lumpy skin disease virus
Rickettsia rickettsii	Hendra virus	Malignant catarrhal fever virus (Alcelaphine herpesvirus type 1)
Saxitoxin	Nipah virus	
Shiga-like ribosome inactivating proteins	Rift Valley fever virus	Menangle virus

(*Continued*)

Table 1.26 (Continued)   Select Biological Agents and Toxins

HHS Select Agents and Toxins	Overlap Select Agents and Toxins	USDA Select Agents and Toxins
South American hemorrhagic fever viruses	Shigatoxin	*Mycoplasma capricolum/ M. F38/M. mycoides Capri*
Flexal		(contagious caprine
Guanarito		pleuropneumonia)
Junin		
Machupo		
Sabia		
Tetrodotoxin	Staphylococcal enterotoxins	*Mycoplasma mycoides mycoides* (contagious bovine pleuropneumonia)
Tickborne encephalitis complex (flavi) viruses	T-2 toxin	Newcastle disease virus (velogenic)
Tetrodotoxin tick-borne encephalitis complex (flavi) viruses		
Central European tick-borne encephalitis		
Far Eastern tick-borne encephalitis		
Kyasanur Forest disease		
Omsk hemorrhagic fever		
Russian Spring and Summer encephalitis		
Variola major virus (smallpox virus) and Variola minor virus (Alastrim)	Venezuelan equine encephalitis virus	Peste des petits ruminants virus
Yersinia pestis (plague)		Rinderpest virus
		Sheep pox virus
		Swine vesicular disease virus
		Vesicular stomatitis virus (exotic)

USDA Plant Protection and Quarantine (PPQ) Select Agents and Toxins

Candidatus Liberobacter africanus	*Ralstonia solanacearum* race 3, biovar 2	*Xanthomonas oryzae* pv. oryzicola
Candidatus Liberobacter asiaticus	*Schlerophthora rayssiae* var zeae	*Xylella fastidiosa* (citrus variegated chlorosis strain)
Peronosclerospora philippinensis	*Synchytrium endobioticum*	

Source: Adapted from CDC. Compiled from U.S. Department of Health and Human Services and U.S. Department of Agriculture 7 CFR Part 331, 9 CFR Part 121 and 42 CFR Part 73 as of February 23, 2006.

The total dose and the dose response are predictors of the harm to an individual. This applies to chemical and biological agents and radiation exposure, as well as hazardous materials and pharmaceuticals.

Half-Life

"Half-life" means the time it takes half of something to undergo a process. It is often applied to a radioactive material to determine how long it will take to decay (Figure 1.21). It can also be applied to other processes, such as biological excretion time or chemical degradation. If the half-life of a substance is one year, it means that after one year from its creation, one-half of the original, active ingredients will remain. In two years, one-fourth of the original amount will remain. In three years, one-eighth will remain.

Radioactivity

Ionizing radiation cannot be sensed by humans, nor can it be detected by field chemistry techniques. Every unknown material should be first screened for ionizing radiation hazards for the safety of the analyzing technician, contamination prevention, and subsequent characterization of the unknown substance.

Some radiation does not have the ability to ionize atoms. Simple radiation is the transfer of heat energy by electromagnetic waves. Radiation occurs without a transfer medium, unlike conduction and convection. The transfer of energy from the sun across nearly empty space is accomplished through radiation. Radiation differs in wavelength, frequency, and energy.

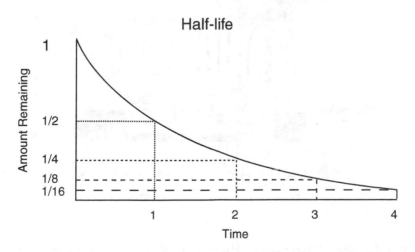

Figure 1.21 Half-life decay and remaining content of an active substance.

"Ionizing radiation" means the waves of radiation have the ability to turn an atom into an ion by removing electrons from the atom (Figure 1.22). When atoms in the human body are ionized by radiation, the affected molecule becomes reactive or is destroyed, causing injury. Random ionization of biological material can produce free radicals, which in turn react with other molecules. Ionizing radiation can break existing chemical bonds or form new bonds or crosslink molecules, making them ineffective. If the ionized

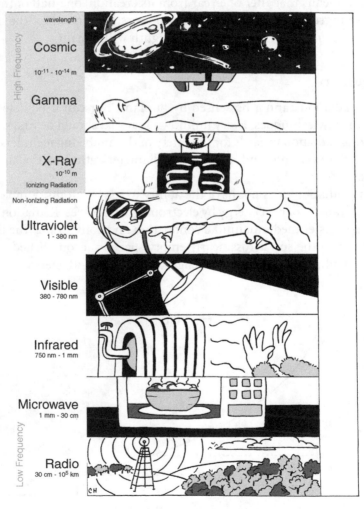

Figure 1.22 Radiation spectrum. Higher energy radiation has the power to ionize molecules.

molecule is connected to genetic function, cancer in the affected individual may occur or a genetic mutation may be passed to future offspring.

Types of Ionizing Radiation

Gamma Rays

Gamma rays are weightless packets of energy with a neutral electrical charge. With no mass and no charge, gamma rays can penetrate six inches of lead or three feet of concrete. Most gamma rays can pass completely through a human, causing ionizing destruction along the way. Gamma rays are emitted from the nucleus of a source atom, and alpha and beta particles may be present. Gamma rays are emitted from substances such as cobalt-60, cesium-137, and uranium.

X-Rays

X-rays have essentially the same properties of gamma rays, but have lower energy. A few millimeters of lead can stop most x-rays. While gamma rays are emitted from the nucleus and are often accompanied by alpha and beta particles, x-rays are produced outside the nucleus. X-rays are most often produced by machines and are unlikely to be encountered through an unknown substance.

Beta Particles

Beta particles are a type of ionizing radiation. Beta particles are high energy, lightweight particles with a negative or positive charge. A negatively charged beta particle is an electron and a positively charged beta particle is a positron. Due to some amount of mass as opposed to no mass, beta particles can be stopped by a few layers of aluminum foil or clothing, although clothing is not necessarily approved personal protective equipment. The range of a beta particle is dependent on its energy and the material it is traveling in. For example; a P-32 beta has a range in air of about 20 feet (7 meters). Beta particles are emitted from substances such as carbon-14, strontium-90, and tritium.

Alpha Particles

Alpha particles consist of two protons and two neutrons and are about 7,000 times more massive than a beta particle. The increased mass means alpha particle radiation can be stopped by a sheet of paper or the dead layer of skin cells on a human. Comparatively, alpha particle radiation is not much of a hazard to the skin, but there are serious inhalation, ingestion, and injection hazards. Epithelial cells that line the respiratory and gastrointestinal tract have no dead layer of cells for protection. A radioisotope dust that is swallowed or inhaled will settle directly on live cells and cause local irradiation injury. The same type of injury can occur with eye contact. Because alpha particles are large and prone to collision while delivering energy over a short distance, they have potential to cause severe biological damage. Alpha radiation can only travel a few centimeters through air. Alpha particles are emitted from substances such as plutonium, radon, and radium.

Neutrons

Neutron radiation is present in background cosmic radiation. It is also produced spontaneously or during a fission reaction. The most significant neutron hazard to be encountered is that which occurs as a result of a nuclear bomb blast or a fission reactor accident. In both cases, other types of radiation would be present.

If neutron radiation alone were to be encountered in the field, it would most likely be from Plutonium-239. Unshielded Plutonium-239 would also emit alpha, beta, and gamma radiation. If Plutonium-239 were suspected inside a heavily shielded container, a neutron detector would be helpful in identifying the source through the sidewall of the container.

Beyond these extreme and limited cases, a neutron detector is not necessary in a radiation hazard screen of unknown substances. Shielding effectiveness is depicted in Figure 1.23.

Radioisotopes and Emissions

A radioisotope is an unstable element that decays or disintegrates spontaneously and emits radiation. Approximately 5,000 natural and artificial radioisotopes have been identified according to the U.S. Nuclear Regulatory Commission. Many of these have extremely short half-life and exist only in research laboratories.

Table 1.27 shows a list of some common radioisotopes and the type of radiation they emit.

Each isotope has a signature that uniquely identifies it to certain detection equipment. The signature involves the type of radiation, wavelength, decay rate, etc. Many of these characteristics are beyond the capabilities of basic field survey equipment. As such, radiation hazard identification will be limited to simply recognizing the presence of radioactive material and putting the hazard into perspective with the entire situation as it is occurring.

Some isotopes occur naturally from the creation of the universe and others are decay products of original isotopes (Table 1.28). Others are man-made (Table 1.29).

Units of Measurement

Several units of measurement are used to quantify radiation energy. Units are established to measure the radioactivity of a source, exposure at a distance from the source, radiation actually absorbed by an object, and equivalent dose when comparing types of radiation (Table 1.30).

Hazards of Sources and Emissions

Radioactive sources have varying characteristics. A source may produce more than one type of emission. It may have a "rapid fire" rate of decay or it may be more sluggish. A source may be concentrated or scattered about an area due to man-made or natural circumstances. Radioactive sources may have accompanying chemical and physical hazards.

Encountering a radioactive material is unlikely due to stringent controls, but there have been a few incidents of inadvertent contamination. This

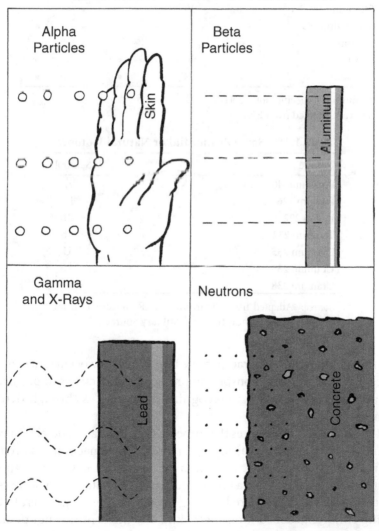

Figure 1.23 Shielding effectiveness is based on type and energy of ionizing radiation.

Table 1.27   Commonly Encountered Radioisotopes

Name	Atomic Number	Alpha	Beta	Gamma
Americium-241	95	•		•
Cesium-137	55		•	•
Cobalt-60	27		•	•
Iodine-129 and Iodine-131	53		•	•
Plutonium	94	•	•	•
Radium	88	•		•
Radon	86	•		
Strontium-90	38		•	
Technetium-99	43		•	•
Tritium*	1		•	
Thorium	90	•		•
Uranium	92	•		•

*Tritium is a specific isotope, H-3.
Source: Adapted from EPA.

Table 1.28   Some Primordial or Natural Isotopes

Isotope	Symbol
Potassium 40	^{40}K
Radium 226	^{226}Ra
Radon 222	^{222}Rn
Thorium 232	^{232}Th
Uranium 235	^{235}U
Uranium 235	^{235}U
Uranium 238	^{238}U

Source: Adapted from Environmental Radioactivity from
Natural, Industrial and Military Sources.

means the identification of a radioactive hazard is a low occurrence but high significance event. If forces conspire to present a radioactive sample for qualitative analysis, the only way to recognize the hazard is through detection equipment.

A 1987 incident underscores the importance of radiation detection equipment and early warning. A gamma ray radiotherapy machine was removed from an abandoned building. Less than an ounce of 137 Cs, with an emission rate of about 1,400 Ci, was removed from its container by looters. The blue glowing powder was shared among people as a novelty. Several people, including children, rubbed it on their skin. The incident was not defined as a radiation emergency for 2 weeks. Of 249 people who were documented as contaminated, 5 died and 49 required hospitalization.

Table 1.29 Some Human Produced Isotopes

Isotope	Symbol
Cesium 137	^{137}Cs
Iodine 129	^{129}I
Iodine 131	^{131}I
Plutonium 239	^{239}Pu
Strontium 90	^{90}Sr
Technetium 99	^{99}Tc
Tritium	^{3}H

Source: Adapted from Environmental Radioactivity from Natural, Industrial and Military Sources.

When a radioactive hazard is identified, its relative risk can often be quantified by the same detection equipment. The radioactive hazard is relative to other hazards of the compound and the area in which it is located.

For example, uranium hexafluoride is radioactive. The U.S. Department of Transportation regulates the shipment of uranium hexafluoride. DOT's Emergency Response Guidebook Guide 166 makes it clear that radiation is not the most significant hazard in a transportation accident involving the release of uranium hexafluoride. Guide 166 states these points to consider as potential health hazards:

- Radiation presents minimal risk to transport workers, emergency response personnel and the public during transportation accidents. Packaging durability increases as potential radiation and criticality hazards of the content increase.
- Chemical hazard greatly exceeds radiation hazard.
- Substance reacts with water and water vapor in air to form toxic and corrosive hydrogen fluoride gas and an extremely irritating and corrosive, white, water-soluble residue.
- If inhaled, may be fatal.
- Direct contact causes burns to skin, eyes, and respiratory tract.
- Low-level radioactive material; very low radiation hazard to people.

Conversely, the intentional dispersion of radioactive iodine in a public area would most likely present a relatively high radiation hazard compared to chemical or physical hazards.

Nerve Agents

Nerve agents are a particular type of chemical agent that poisons the ability of nerves to conduct impulses properly to other nerves or target tissues.

Table 1.30   Radiation Units of Measurement

Measurement Use	United States Units	International System Units
Measurement of the radioactivity of a source.	A Curie (Ci) is equal to the amount of a radioactive isotope that decays at the rate of 3.7×10^{10} disintegrations per second which is the number of disintegrations observed from 1 g of radium. Often radioactivity is expressed in smaller units such as thousandths (mCi), millionths (uCi), or billionths (nCi) of a curie. The Becquerel (Bq) is equal to the quantity of radioactive material that will have one transformation in one second. Often radioactivity is expressed in larger units such as thousands (kBq), millions (MBq) or billions (GBq) of becquerels. One curie equals 3.7×10^{10} Bq.	
A direct measurement of exposure. As distance from the radioactive source increases, exposure to a detector decreases. It is a convenient measurement, but it is limited to gamma and x-rays in air.	A Roentgen (R) is a unit of radiation exposure equal to the quantity of ionizing radiation that will produce one electrostatic unit of electricity in one cubic centimeter of dry air at 0°C and standard atmospheric pressure. The rate of radiation is measured over time in roentgens per hour (R/hr) or milliroentgens per hour (mR/hr). Roentgen applies only to gamma and x-rays, not alpha, beta, or neutron radiation.	A Roentgen (R) is a unit of radiation exposure equal to the quantity of ionizing radiation that will produce one electrostatic unit of electricity in one cubic centimeter of dry air at 0°C and standard atmospheric pressure. The rate of radiation is measured over time in roentgens per hour (R/hr) or milliroentgens per hour (mR/hr).
Measurement of the radiation absorbed by an object. It does not count the radiation that passes through the material and is not absorbed. This unit is for use on inanimate objects because it does not describe the biological effect of the radiation.	A rad (radiation absorbed dose) is a measurement of radiation absorbed by any material in the amount of 100 ergs (mJ/kg) per gram. It is related to the material being measured. Injury is dependent on the type of radiation. 10 rad of x-ray does not cause the same injury as 10 rad of alpha radiation.	A Gray (Gy) is the unit for the energy absorbed from ionizing radiation, equal to one joule per kilogram. It is related to the material being measured.
Measurement of an equivalent dose of radiation. Use this measurement to compare biological radiation exposures. These units are applied to all forms of ionizing radiation and consider the type of radiation.	A rem (Roentgen equivalent man) is the exposure required to produce the same biological effect as one rad of x-rays or gamma rays. Rem is rad multiplied by a relative biological effectiveness (RBE) factor and considers the type of radiation as well as the radiation absorbed dose. Rem is used to describe the ability to cause biological damage. Since these are relatively large measurements, equivalent dose is often expressed in terms of thousandths of a rem, or a millirem.	One Sievert (Sv) is equal to 100 rems. Equivalent dose is often expressed in terms of millionths of a Sievert, or a micro-Sievert.

Nerve agents are stable, easily dispersed, and highly toxic. They can be manufactured relatively easily and have rapid effects when absorbed through the skin or respiratory system.

The mechanism of action involves inhibition of the body's ability to deactivate a natural neurotransmitter called acetylcholine; other neurotransmitters can be affected. Acetylcholine is one of several chemicals in the body secreted by one nerve cell to another in order to propagate a signal through the nervous system. If acetylcholine was left unchecked in the space between two nerve cells, the signal would not stop being sent and thus the output, such as a muscle contraction, would not stop.

The enzyme (a biological catalyst) responsible for deactivating acetylcholine is acetylcholinesterase. Nerve agents act by binding to acetylcholinesterase, inactivating it, and then allowing acetylcholine to continue to stimulate certain nerve functions (Figure 1.24).

Role of Nerve Agent in Nerve Transmission

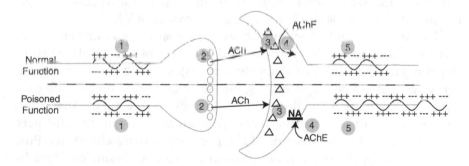

Normal Nerve Impulse Transmission (Top)

1. Depolarization wave moves down pre-synaptic nerve sheath.
2. Depolarization wave stimulates vesicles to release neurotransmitter acetylcholine (ACh).
3. After quickly crossing synapse, ACh plugs into post-synaptic receptor and stimulates depolarization down nerve sheath.
4. Enzyme acetylcholinesterase (AChE) immediately inactivates ACh and depoolarization stimulus stops.
5. A normal "on-off burst" of depolarization carries the message to the end of the nerve.

Poisoned Nerve Impulse Transmission (Bottom)

1. Depolarization wave moves down pre-synaptic nerve sheath.
2. Depolarization wave stimulates vesicles to release neurotransmitter acetylcholine (ACh).
3. After quickly crossing synapse, ACh plugs into post-synaptic receptor and stimulates depolarization down nerve sheath.
4. Enzyme acetylcholinesterase (AChE) is chemically blocked by nerve agent. Depoolarization stimulus does not stop.
5. A constant "on" message is transmitted to the end of the nerve. The receiving organ is stimulated to function without rest. Initial reaction is overstimulus of the organ followed by fatigue.

Figure 1.24 Depiction of a nerve synapse and inhibition of acetylcholinesterase.

Other compounds have the ability to inhibit acetylcholinesterase such as organophosphate and carbamate insecticides; however, nerve agents are especially potent inhibitors.

Acetylcholine is present in both the central and peripheral nervous systems, and it may be used to increase or decrease the activity of an organ. Another neurotransmitter may be used to counter the effect of the acetylcholine containing system so that a balance is attained and function controlled. The body uses several other compounds to transmit messages. Chemicals like epinephrine, serotonin, dopamine, and γ-aminobutyric acid (GABA) all work in concert to produce a response based on the needs of the person. When acetylcholine floods the system unabated, the body loses control of essential systems and dies.

Acetylcholinesterase inhibiting material can also have long term effects. Sublethal doses can chronically affect central nervous system function, causing several mental health consequences.

A very small amount of nerve agent is a lethal dose. A 10 microliter drop of VX nerve agent (enough to cover the ear of President Roosevelt's bust on a dime) is a lethal dose when absorbed through the skin. The recently declassified novichok series may be 5–10 times more toxic than VX.

Nerve agents tend to act quickly and their effects may or may not be reversible with an antidote. An inhaled nerve agent will be absorbed by target organs quickly. Skin absorption may delay for 20–30 minutes the nerve agent arrival at target organs, but death can occur quickly after the onset of symptoms if the dose is high enough. Nerve agents that affect humans also can kill insects, birds, wild animals, livestock, and pets. These signs can be an indicator of the presence of these chemicals. Pure nerve agents appear as clear, oily substances. Contaminants or degradation products may produce a darker appearance. Dispersed nerve agent is not likely to be visible.

Nerve agents may be inhaled, injected, ingested, or absorbed. They may be dispersed in pure form or in solvent, on dust and fibers, thickened, or enhanced in other ways. Altered forms of nerve agents may make them undetectable by air monitors.

The more common nerve agents are usually referred to by their military designations. A few are listed in Table 1.31 by properties. Note that the only volatile nerve agent shown is GB, or sarin. GB is a respiratory hazard. The others are skin contact hazards because their volatility is low and they produce little vapor. All nerve agents are respiratory hazards if dispersed in air as a mist or vapor; inhalation hazard decreases as the agent condenses on surface and the skin contact hazard remains. Many materials will inhibit cholinesterase. Molecular structure varies as does elemental content of the substance as shown in Table 1.32.

Table 1.31 Physical Properties of Common Nerve Agents at 77°F

	GA (Tabun)	GB (Sarin)	GD (Soman)	GF (Cyclosarin)	VX	Water
Density, g/m^3	1.073	1.089	1.022	1.120	1.008	1
Boiling point, °F	464	316	388	462	568	212
Freezing point, °F	18	−69	−44	−22	<−60	32
Vapor pressure, mmHg	0.07	2.9	0.4	60.0	0.0007	23.756
Water solubility, %	10	Miscible	2	2	Slight	Miscible

Table 1.32 Elemental Content of Nerve Agents

Nerve Agent	C	H	O	P	F	S	N	CN	Cl
DMAEDMAFP (GP)	•	•	•	•	•	•			
DMAPDMAFP (EA)	•	•	•	•	•	•			
G-Series	•	•	•	•	Some	Some	Some	Some	Some
Sarin (GB)	•	•	•	•	•				
Soman (GD)	•	•	•	•	•				
Tabun (GA)	•	•	•	•			•		•
V Series	•	•	•	•		•			
V-Gas (Vx)	•	•	•	•		•	•		
VM (VM)	•	•	•	•		•	•		
VX (VX)	•	•	•	•		•	•		
GV-Series	•	•	•	•	•		•		
A-230	•	•	•	•	•		•		•
A-232	•	•	•	•	•		•		•
A-234	•	•	•	•	•		•		•

Source: Adapted from *Handbook of Chemical and Biological Warfare Agents* and *Novichoks: The Dangerous Fourth Generation of Chemical Weapons.*

Blood Agents

"Blood agent" is a military designation of a group of chemical warfare agents that disrupt the ability of the blood system to perfuse the body. Three types of blood agents are:

- COX (cytochrome oxidase) inhibiting blood agents. These agents prevent the transfer of oxygen from the blood cell to tissues. COX inhibiting agents include cyanides, halogenated cyanides, and hydrogen sulfide. These materials are in common commercial use and could be adapted for terror.

- Arsine blood agents. Arsine destroys red blood cells and distrupts vital organ function. Arsine blood agents include arsine and materials that readily decompose to produce arsine.
- Carbon monoxide blood agents. Carbon monoxide binds to hemoglobin in the red blood cell, poisoning the bonding site used to transport oxygen. Carbon monoxide blood agents include carbon monoxide and materials that readily decompose to produce carbon monoxide.

Vesicants (Blister Agents)

Vesicants are also known as "blister agents" (Table 1.33). Vesicants are extremely destructive to tissue and cause irritation, blistering, and eventual death of tissue. Vesicants are highly reactive chemicals that combine with proteins, DNA, and other cellular components to result in cellular changes immediately after exposure.

Vesicants were the most commonly used chemical warfare agents during World War I. These chemical warfare agents are based on sulfur, nitrogen, or arsenic. Sulfur and nitrogen mustards are more stable than arsenical vesicants. Although tightly restricted due to the ban on chemical warfare agents, some of these compounds are used in medical research and have shown promise in cancer treatment.

Most blister agents are relatively persistent and are readily absorbed by all parts of the body. These vesicants cause inflammation, blisters, and general destruction of tissues.

The most likely routes of exposure are inhalation, skin contact, and eye contact. Inhalation exposure produces irritation of the respiratory tract and difficulty breathing. Skin and eye contact result in irritation, chemical burns, and blistering. Ingestion of vesicant material, such as swallowing contaminated mucous, can cause nausea and vomiting. Clinical effects may occur immediately or may be delayed for 2 to 24 hours depending on the compound. A delayed effect is a characteristic of mustard agent. There are 13 vesicant agents that are considered chemical warfare agents. Other compounds and mixtures exist that are irritating and can cause blistering. Elemental content of blister agents is shown in Table 1.34.

Vesicants tend to have low vapor pressure and if dispersed in the air as a weapon are contacted initially by inhalation of vapor and later by skin contact of the condensate once the air has cleared.

Riot Control Agents (Irritants)

Irritants are chemicals capable of destroying human tissue and may be lethal if exposure is extended beyond a short time. The U.S. Department of Defense

Table 1.33 Physical Properties of Common Blister Agents

	HD (Sulfur Mustard)	HN-1 (Nitrogen Mustard)	HN-2 (Nitrogen Mustard)	HN-3 (Nitrogen Mustard)	L (Lewisite; Arsenical)	Water
Density, g/m^3	1.27 @ 68°F	1.09 @ 77	1.15 @ 68	1.24 @ 77	1.89 @ 68	1 @ 77
Boiling point, °F	421	381	167 @ 15 mmHg	493	374	212
Freezing point, °F	58	-61.2	-85	-26.7	32.18-64.4	32
Vapor pressure, mmHg	0.072 @ 68°F	0.24 @ 77°F	0.29 @ 68°F	0.0109 @ 77°F	0.394 @ 68°F	23.756 @ 77°F
Water solubility, %	< 1%	Slight	Slight	Insoluble	Insoluble	Miscible

Table 1.34 Elemental Content of Blister Agents

Blister Agent	Element									
---	C	H	O	P	F	S	N	CN	Cl	As
Dimethylsulfate (no designation)	•	•	•			•			•	
Ethyldichloroarsine (ED)	•	•							•	•
Lewisite (L)	•	•							•	•
Methyldichloroarsine (MD)	•	•							•	•
Mustard (H or HD)	•	•				•			•	
Nitrogen mustard-1 (HN-1)	•	•					•		•	
Nitrogen mustard-2 (HN-2)	•	•					•		•	
Nitrogen mustard-3 (HN-3)	•	•					•		•	
O-mustard (T)	•	•	•			•			•	
Phenyldichloroarsine (PD)	•	•							•	•
Phosgene oxime (CX)	•	•	•				•		•	
Sesqui-Mustard (Q)	•	•				•			•	

Source: Adapted from *Handbook of Chemical and Biological Warfare Agents.*

defines an irritant as "any chemical, that is not listed in the Chemical Weapons Convention, which can produce rapidly in humans sensory irritatation or disabling physical effects which disappear within a short time following termination of exposure" (Table 1.35).

Riot control agents are divided into lachrymators and vomiting agents, although they cause more than one sign or symptom such as tearing, nausea, vomiting, a bad taste, pulmonary edema, sneezing, coughing, or a generalized bad feeling.

Table 1.35 Examples of Some Irritants

Lachrymators	Designator
Orthochlorobenzylidene malononitrile	CS
Dibenzoxazepine	CR
Bromobenzyl cyanide	CA
Bromoacetone	BA
Dibenzoxazepinem (suspected carcinogen)	CR
Vomiting Agents	
Diphenylchlorarsine	DA
Diphenylcyanoarsine	DC
Diphenylaminearsine chloride (Adamsite)	DM

Source: Adapted from North American Treaty Organization (NATO).

Riot control agents are solids that are released as a fine powder or aero-solized in a solvent. They have low toxicity and can produce sensory irritation or disabling physical effects that disappear soon after exposure ceases.

Pepper spray, sometimes called OC, contains ground peppers and extract. The active ingredient is capsaicin. Pepper spray is irritating, but not a military-grade irritant.

Toxic Industrial Material

Toxic industrial materials are substances other than chemical warfare agents that have harmful effects on humans. Chemicals that may have military significance may also have industrial significance, such as chlorine gas and hydrogen cyanide. Toxic industrial materials (TIMs) are sometimes called toxic industrial chemicals (TICs). A TIM is a material that has an LCt50 (lethal concentration for 50% of the population multiplied by exposure time) less than 100,000 mg-min/m^3 in any mammalian species, and is produced in quantities exceeding 30 tons per year at one production facility. TIMs could be used to make a significant terrorist impact, so factors other than lethality are used to define TIMs.

TIMs are listed by hazard index ranking in Table 1.36. High hazard ranking indicates a widely produced, stored, or transported TIM that has high toxicity and is easily vaporized. Medium hazard ranking indicates a TIM that may rank high in some categories and lower in others. A low hazard ranking indicates a TIM that is not likely to be a hazard unless specific situations exist.

Persistence

Persistence describes the ability of a chemical warfare agent to remain in place. It is used to help describe the duration of potency.

Persistence can be calculated and listed in references. An agent will most likely dissipate in the air or undergo degradation in place due to environmental or decontamination factors.

Factors that affect the dissipation of an agent include ambient temperature and the vapor pressure and molecular weight of the agent.

Factors that affect the degradation of an agent in place include temperature, humidity, and exposure to various decontamination agents. The contaminated material also affects persistency. For example, a chemical warfare agent may dissipate through a material more thoroughly than a decontamination agent. The surface of the material may test as decontaminated but several hours or days later the chemical warfare agent may diffuse through the material and reappear on the surface as a toxic hazard.

Table 1.36   TIMs Listed by Hazard Index Ranking

High	Medium	Low
Ammonia	Acetone cyanohydrins	Allyl isothiocyanate
Arsine	Acrolein	Arsenic trichloride
Boron trichloride	Acrylonitrile	Bromine
Boron trifluoride	Allyl alcohol	Bromine chloride
Carbon disulfide	Allylamine	Bromine pentafluoride
Chlorine	Allyl chlorocarbonate	Bromine trifluoride
Diborane	Boron tribromide	Carbonyl fluoride
Ethylene oxide	Carbon monoxide	Chlorine pentafluoride
Fluorine	Carbonyl sulfide	Chlorine trifluoride
Formaldehyde	Chloroacetone	Chloroacetaldehyde
Hydrogen bromide	Chloroacetonitrile	Chloroacetyl chloride
Hydrogen chloride	Chlorosulfonic acid	Crotonaldehyde
Hydrogen cyanide	Diketene	Cyanogen chloride
Hydrogen fluoride	1,2-Dimethylhydrazine	Dimethyl sulfate
Hydrogen sulfide	Ethylene dibromide	Diphenylmethane-4,4'-diisocyanate
Nitric acid, fuming	Hydrogen selenide	Ethyl chloroformate
Phosgene	Methanesulfonyl chloride	Ethyl chlorothioformate
Phosphorus trichloride	Methyl bromide	Ethyl phosphonothioic dichloride
Sulfur dioxide	Methyl chloroformate	Ethyl phosphonic dichloride
Sulfuric acid	Methyl chlorosilane	Ethyleneimine
Tungsten hexafluoride	Methyl hydrazine	Hexachlorocyclopentadiene
	Methyl isocyanate	Hydrogen iodide
	Methyl mercaptan	Iron pentacarbonyl
	Nitrogen dioxide	Isobutyl chloroformate
	Phosphine	
	Phosphorus oxychloride	Isopropyl chloroformate
	Phosphorus pentafluoride	Isopropyl isocyanate
	Selenium hexafluoride	n-Butyl chloroformate
	Silicon tetrafluoride	n-Butyl isocyanate
	Stibine	Nitric oxide
	Sulfur trioxide	n-Propyl chloroformate
	Sulfuryl chloride	Parathion
	Sulfuryl fluoride	Perchloromethyl mercaptan
	Tellurium hexafluoride	sec-Butyl chloroformate
	n-Octyl mercaptan	tert-Butyl isocyanate
	Titanium tetrachloride	Tetraethyl lead
	Trichloroacetyl chloride	Tetraethyl pyrophosphate
	Trifluoroacetyl chloride	Tetramethyl lead
	Toluene 2,6-diisocyanate	Toluene 2,4-diisocyanate

Source: Reproduced from International Task Force 25: Hazard from Industrial Chemicals
Final Report, April 1998.

Persistence is a relative term. For example, nerve agent VX would be considered a persistent agent due to its high molecular weight of 267, its low vapor pressure of 0.0007 mmHg, and its stability. Nerve agent GB (sarin) would be considered much less persistent agent due to a lighter molecular weight of 140, a vapor pressure of 2.9 mmHg at 77° F, and its solubility in water as well as its susceptibility to degradation in the presence of water vapor over a week or so.

Signs of a WMD Attack

A chemical incident is characterized by a rapid onset of medical symptoms (minutes to hours) and can have observed signatures such as colored residue, dead foliage, pungent odor, and dead insect and animal life.

With biological incidents, the onset of symptoms usually requires days to weeks, and there are typically no characteristic signatures because biological agents are usually odorless and colorless. The area affected can be greater due to the migration of infected individuals because of the delayed onset of symptoms. An infected person could transmit the disease to another person.

An incident involving radioactivity may produce symptoms immediately if the dose is high. Signs and symptoms could be delayed if the dose is low. Like biological incidents, radioactive incidents could spread if the source migrates from the original area. Radioactivity cannot be sensed by humans and must be detected by a radiation monitor.

The overall importance of recognizing an unfolding WMD attack cannot be overstated, but the details of recognition are well beyond the scope of this book. If not familiar with the signs of a WMD attack, you are encouraged to become more familiar through other sources such as "Guidelines for Responding to a Chemical Weapons Incident," a document developed by the Domestic Preparedness Program.

References

2004 Emergency Response Guidebook, U.S. Department of Transportation, Washington, DC, 2004.

Byrnes, Andrew, DuPont, Hank, Matthew, David, McCartt, Jack, Noll, Gregory, G., and Yoder, Wayne, *The Jack Rabbit II Project's Impacts on Emergency Responders Catastrophic Releases of Liquefied Compressed Chlorine*, U.S. Army Dugway Proving Ground, Dugway, UT, September 30, 2017.

Cooper, P.W., *Explosives Engineering*, Wiley-VCH, New York, 1996.

Departments of the Army, the Navy, and the Air Force. *NATO Handbook on the Medical Aspects of NBC Defensive Operations AMedP-6(B)*, Washington, DC, 1996.

Dominique, M., http://encyclopedia.airliquide.com, AIR LIQUIDE S.A., Paris, 2006.

Donev, J., Afework, B., and Hanania, J, (2018). *Energy Education: Phase Change*, https://energyeducation.ca/encyclopedia/Phase_change [accessed: February 4, 2020].

Eisenbud, M. and Gesell, T., *Environmental Radioactivity from Natural, Industrial, and Military Sources*, 4th ed., Academic Press, San Diego, 1997, 135.

Ellison, D.H., *Handbook of Chemical and Biological Warfare Agents*, CRC Press, Boca Raton, FL, 1999.

Environmental Protection Agency. *How to Prevent Runaway Reactions – Case Study: Phenol-Formaldehyde Reaction Hazards*, Environmental Protection Agency, Washington, DC, August, 1999.

Environmental Protection Agency. *Title 40 Code of Federal Regulations*, Part 261.23, Environmental Protection Agency, Washington, DC, June 26, 2006.

Jennings, M., *HHS and USDA Select Agents and Toxins*, Centers for Disease and Control, Department of Health and Human Services, Washington, DC, February 23, 2006.

National Fire Protection Association, *NFPA 30 Flammable and Combustible Liquids Code*, 2002 ed., NFPA, Quincy, MA, 2003.

National Fire Protection Association, *NFPA 472 Competence of Responders to Hazardous Materials/Weapons of Mass Destruction Incidents*, 2018 ed., NFPA, Quincy, MA, 2002.

National Institute for Occupational Safety and Health, *NIOSH Pocket Guide to Chemical Hazards*, Superintendent of Documents, Centers for Disease Control and Prevention, Government Printing Office, Washington, DC, September 2005.

Rickens, S., *Media Release*, Commonwealth of Pennsylvania Department of Environmental Protection, Harrisburg, July 12, 2006.

Stuempfle, A. K., Howells, D. J., Armour, S. J., Boulet, C. A., and Aberdeen Proving Ground, MD. *International Task Force 25: Hazard from Industrial Chemicals Final Report*, Edgewood Research Development and Engineering Center, Aberdeen Proving Ground, MD, April 1998.

The American Heritage® *Dictionary of the English Language*, 4th ed., Houghton Mifflin, Boston, 2000.

U.S. Army Soldier and Biological Chemical Command (SBCCOM). *Guidelines for Responding to a Chemical Weapons Incident*, Publication L126/QRG-C, Domestic Preparedness Program, Washington, DC, August 1, 2003.

U.S. Department of Energy. *Primer on Spontaneous Heating and Pyrophoricity*, United States Department of Energy, Washington, DC, December 1, 1994. *The Radiological Accident in Goiania*, International Atomic Energy Agency, Vienna, 1989.

U.S. Environmental Protection Agency. *Radionuclides*, U.S. Environmental Protection Agency, Washington, DC, December 14, 2005.

United States Department of Labor. *Title 29 Code of Federal Regulations*, Part 1910.106, United States Department of Labor, Washington, DC, retrieved September 26, 2006.

United States Department of Transportation. *Title 49 Code of Federal Regulations*, Part 173, United States Department of Transportation, Washington, DC, October 1, 2004.

United States Environmental Protection Agency, *Prevention of Reactive Chemical Explosions Case Study: Waste/Oxidizer Reaction Hazards*, Washington, DC, April 2002.

Weast, R.C., *Handbook of Chemistry and Physics*, 57th ed., CRC Press, Boca Raton, FL, 1976.

Hazards

2

Figure 2.1 It seems that hazards are just a way of life to some people. Many people are injured or killed because they did not understand the hazards presented by a situation.

You can be injured or killed by an unknown substance due to its hazards. Risk includes injury due to physical form, radioactivity, reactivity, corrosivity, flammability, toxicity, and explosive properties of the material. This chapter describes the hazards and secondary hazards that might occur. Understanding these hazards will help you recognize and characterize hazards during analysis. When hazards have been characterized, you can predict how an unknown material will react on a larger scale. This systematic use of information as it becomes available will help you determine the events that may occur and the magnitude of their consequences (Figure 2.1).

Risk analysis is a procedure to identify threats and vulnerabilities, analyze them to ascertain the exposures, and highlight how the impact can be eliminated or reduced. If the unknown material can be identified with a high degree of confidence, many resources become available for the risk analysis. However, identification is not a prerequisite for risk analysis if hazard characteristics are known. The risk analysis you develop will be a fluid concept that changes with the circumstances of the situation.

Physical Form

Unknown material can present hazards due to physical form of the material. Physical form includes the phase, size of particle, mobility, and temperature. Noting these general characteristics is useful in predicting behavior and identifying hazards.

Temperature

A sudden release of pressure on a gas will cause a drop in temperature. This is what causes a carbon dioxide fire extinguisher to form frost on the cylinder after discharge. Any gas container that has experienced a sudden depressurization will be colder than it was before the release. The sudden drop in temperature can be enough to cause a phase change, such as the dry ice formed when carbon dioxide is discharged too close to an object. The dry ice will slowly warm to a gas and dissipate over time, but if you touch the dry ice, which is extremely cold, you will incur frostbite injury.

Material in a closed container with increasing pressure will experience a rise in temperature. The container could be pressurized from an external source, such as a high-pressure air pump used to fill a firefighter's self-contained breathing apparatus cylinder. The pressure could also develop from a reaction taking place within the container. In either case, the temperature of the container and its contents will rise and could pose a thermal hazard.

Any time a phase change is possible, you should be aware of how the phase may impact your ability to control the material. A substance changing phase can present a mobility hazard – that is, the material may be able to escape from its container or the area in which you are working.

Solids

Solids tend to be the least hazardous phase due to decreased mobility; these materials generally are not mobile and are less likely to be encountered unexpectedly. However, solid material tends to be more concentrated than other phases. Additionally, solid phase materials can be inhalation and fire or explosive hazards if finely divided.

When solids are finely divided into powder form, they become more mobile. Relatively inert material, such as wheat flour, can become explosive if finely divided and dispersed through the air. In order to prevent fire and explosion, flour is moved between containers in a manner similar to flammable liquids – i.e., bonding and grounding cables are used, and ignition sources are eliminated. Relatively high concentrations of solid combustible material in air are necessary to produce fire or explosive conditions.

Solid particle sizes of less than about 5 microns will greatly affect the mobility of the material (Figure 2.2). These small particles are able to float in the air for extended periods, being held buoyant by air molecules. *Bacillus anthracis* spores are about 1 × 1.5 microns. If static charge is eliminated, these spores become a long lasting inhalation hazard. The same is true of finely divided silica, asbestos, and any other solid.

Solid hazardous material will maintain its chemical characteristics. For example, powdered sodium hydroxide in the air is a respiratory and skin

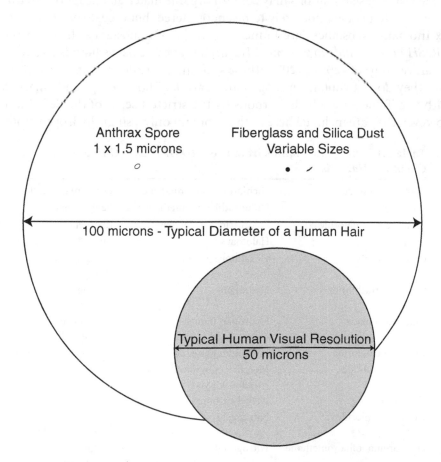

Figure 2.2 Particle size of solid material relative to visual resolution.

contact hazard. Solids that are heated could transition to liquid phase and become more mobile. Solids that absorb water can become fluid.

Solids in general do not develop significant vapor pressure. A few solids, such as naphthalene flakes (VP 0.08 mmHg, PEL 10 ppm, IDLH 250 ppm), do generate vapors, which can be inhalation and/or fire hazards.

Liquids

Liquids tend to be the most hazardous phase due to increased concentration, mobility, and vapor production. Liquid phase material can be as concentrated as solid material and indeed is often concentrated to decrease shipping cost. Liquids will flow to low areas and pool. Liquids also tend to produce more vapors, increasing their mobility into the air. Generally, liquid phase material presents inhalation hazards due to likely vapor production. High vapor pressure liquids are likely to be fire or explosion hazards. Many liquids are acidic or caustic.

Liquids are likely to be either water-based or organic (burnable) material. Some water-based solutions may contain organic material. Exceptions to this general statement are not likely to be encountered, but they do exist. Thirty-six inorganic substances are found within the 385 substances listed in the *NIOSH Pocket Guide to Chemical Hazards* that can exist as liquids in a temperature range of −20° to 120°F. These substances are defined as inorganic in that they do not contain hydrogen and carbon. Those compounds marked with * in Table 2.1 are hydrocarbons by the strictest sense of the definition, however, they are included here as they contain only a single hydrogen atom,

Table 2.1   Inorganic Liquids from the *NIOSH Pocket Guide to Chemical Hazards*

Boron tribromide	Dichlorotetrafluoroethane	Osmium tetroxide
Bromine	Difluorodibromomethane	Pentaborane
Bromine pentafluoride	Fluorotrichloromethane	Phosgene
Bromoform*	Halothane*	Phosphoric acid
Carbon disulfide	Hexafluoroacetone	Phosphorus (yellow)
Carbon tetrachloride	Hydrazine	Phosphorus oxychloride
Chlorine dioxide	Hydrogen fluoride	Phosphorus trichloride
Chlorine trifluoride	Hydrogen peroxide	Stibine
Chromyl chloride	Mercury compounds (except organo-alkyls)	Sulfur dioxide
Cyanogen	Nickel carbonyl	Sulfuric acid
Cyanogen chloride	Nitric acid	Sulfur monochloride
Dichloromonofluoromethane	Nitrogen dioxide	Sulfur pentafluoride

which is insignificant in relation to the rest of the molecule for the purpose of identifying hazardous characteristics. Significantly, most of these liquids have melting points or boiling points within the temperature range of −20° to 120°F, which would be characteristic of a relatively pure sample.

Generally, water-based liquids are not flammable, but organic liquids are likely to be combustible or flammable and can be ignited easily or be forced to burn. Care must be taken when transferring flammable liquids to avoid static charge or other ignition sources. Some characteristic trends emerge among flammable liquids. The simpler the hydrocarbon molecule, the lower its boiling point and flash point and the higher its vapor pressure.

All liquids generate vapors. Vapor pressure is a rough indicator of inhalation hazard and possible flammability. Water has a vapor pressure of about 18 mmHg at room temperature, and this is a useful benchmark. Organic liquids that evaporate quickly have less surface tension and will not bead up as much as water. This will make it easier to observe the edge of the liquid for evaporation from a flat surface. You can also compare the evaporation rate of the sample to a known liquid, such as acetone (VP 180 mmHg), and extrapolate a vapor pressure estimate. Vapor pressure is a good indicator of relative flammability but a poor predictor of toxicity.

Liquid hazardous material will maintain its chemical characteristics. For example, sulfuric acid liquid or mist in the air is a skin contact hazard. Due to its low vapor pressure, liquid sulfuric acid that is not in the air cannot pose an inhalation hazard. But if sulfuric acid is heated, mixed with reactive material, or is misted into the air from a pressurized leak, the vapor becomes mobile in air and a severe respiratory and skin contact hazard.

Many liquids are corrosive. Nearly all the industrial acids are liquids. Industrial bases can be solids mixed into solutions before use. Liquids that do not contain water are most likely organic and the low viscosity organics are flammable; the high viscosity organics will burn if heated. Liquids are highly mobile and will form gas or vapor proportional to their vapor pressure.

Gases

Gas phase materials are highly mobile and likely to dissipate quickly. Some gases will permeate solid or liquid material and become less mobile, but for the most part, dissipation occurs spontaneously. Gases can be the most difficult to manage once out of the container. Generally, gas phase material presents an inhalation hazard due to highly concentrated gas. Gases that are combustible also present a fire or explosion hazard.

For example, hydrogen cyanide is a gas stored in cylinders. If released into a lab, it becomes an inhalation hazard and is immediately dangerous to life and health at 50 ppm. High toxicity and high mobility make this a dangerous situation. Near the release it also becomes a fire and explosion hazard if the concentration reaches 5.6%.

An unidentified gas is highly mobile and might explode and burn, or poison or suffocate you. Gases and gas contaminated air should always first be monitored for flammability with a combustible gas indicator in conjunction with oxygen concentration. Gases can be tested for a relatively high concentration of corrosive, caustic, oxidizer, and some toxins with wet test strips. Colorimetric air monitoring tubes can be used in a screen to analyze a gas qualitatively. Colorimetric air monitoring screens are available, such as Deluxe Haz Mat III (Sensidyne®) and Simultaneous Test Sets (Drager®). Air monitors using electrochemical and other sensors may detect a gas, but all of the sensor cross sensitivities must be considered in the case of a truly unknown gas. Advanced technology might be useful to identify a gas, but limitations exist.

Odors and Odor Threshold

Human and canine senses can be used as clues to verify the identity of a material. Obviously, exposure to a hazardous material should never be intentional, but descriptions from people inadvertently exposed may be useful in confirming a substance.

Odor cannot be relied on to prevent overexposure. Human sensitivity to odors varies widely. Some chemicals cannot be smelled at toxic concentrations or sometimes cannot be smelled at all and the minimum threshold at which any random person can detect it may vary widely. Some odors can be masked by other odors and some compounds can rapidly deaden the sense of smell.

Because odor is difficult to define and describe in a manner that would be helpful, the characteristics are purposefully vague. The NIOSH Pocket Guide to Chemical Hazards contains a heading called "Physical Description" for each listing, which very briefly describes the appearance and odor of each substance, whether it can be shipped as a liquefied compressed gas, and whether it is used as a pesticide. For example, 2,4-D (2,4-dichlorophenoxyacetic acid) is described as "white to yellow, crystalline, odorless powder [herbicide]." However, most people would recognize it in its more common form as a brown water solution (<1%) with other ingredients and a recognizable odor that is used to kill dandelions and other broadleaf weeds on a lawn, such as Ortho Weed-B-Gon MAX Weed Killer.

Many odors are recognizable by humans, but often in controlled settings. The perception of odors is highly subjective and should not be relied upon as a defining test; rather, odor characteristics of unknown material should be used as suggestive evidence as to the identification of one or more components in a compound. Never intentionally smell a sample. Use odor clues as they may be reported to you by people who may have inadvertently been exposed.

Radiation

Ionizing radiation cannot be sensed by humans, nor can it be detected by field chemistry techniques. You could receive a fatal dose of radiation and not be aware of it at the time.

When atoms in the human body are ionized by radiation, the affected molecule becomes reactive or is destroyed, causing injury. Individual cells may be killed or biochemical messengers such as hormones and neurotransmitters are altered. Radiation may produce toxic substances from existing body chemicals, such as water reacting to form hydrogen peroxide. Damage may be induced primarily by radiation or secondarily by the production of chemical toxins.

Radiation injury may cause death, radiation sickness, cancer, or mutant changes that may be passed onto the next generation via altered genetic material.

Radioactive sources have varying characteristics. A source may produce more than one type of emission. It may have a "rapid fire" rate of decay or it may be slower. A source may be concentrated or scattered about an area due to man-made or natural circumstances. Radioactive sources may have accompanying chemical and physical hazards.

Encountering radioactive material is unlikely due to stringent controls. This means the recognition of a radioactive hazard in an emergency, whether accidental or intentional, is a low occurrence but high significance event. If forces conspire to present a radioactive sample for qualitative analysis, the only way to recognize the hazard is through detection equipment.

When a radioactive hazard is identified, its relative risk can often be quantified by the same detection equipment. The radioactive hazard is relative to other hazards of the compound and the area in which it is located.

For example, uranium hexafluoride is radioactive. The United States Department of Transportation regulates the shipment of uranium hexafluoride. DOT's *Emergency Response Guidebook Guide 166* makes it clear that radiation is not the most significant hazard in a transportation accident involving the release of uranium hexafluoride. Guide 166 states these points to consider as potential health hazards:

- Radiation presents minimal risk to transport workers, emergency response personnel, and the public during transportation accidents. Packaging durability increases as potential radiation and criticality hazards of the content increase.
- Chemical hazard greatly exceeds radiation hazard.
- Substance reacts with water and water vapor in air to form toxic and corrosive hydrogen fluoride gas and an extremely irritating and corrosive white-colored, water-soluble residue.
- If inhaled, may be fatal.

- Direct contact causes burns to skin, eyes, and respiratory tract.
- Low-level radioactive material; very low radiation hazard to people.
- Runoff from control of cargo fire may cause low-level pollution.

Conversely, the intentional dispersion of radioactive iodine in a public area would most likely present a relatively high radiation hazard compared to chemical or physical hazards. The iodine itself would not cause much chemical injury. Radioactive iodine would bind to certain hormones and collect in an exposed person's thyroid. This higher concentration of radioactive iodine in one spot in the body would result in an amplified exposure to one organ compared to a general exposure absorbed by the entire body.

Radiation can only be detected by instrumentation. Without proper protection, you could be lethally injured by radiation before realizing sensation of the exposure.

Reactivity

Air Reactivity

Air-reactive and pyrophoric substances will self-heat, self-react, or spontaneously burn when in contact with air. Air-reactive material may cause secondary hazards:

- Heat
- Fire
- Gas
  - Toxic
  - Corrosive
  - Oxidizing
  - Explosive
- Corrosive residue

Air-reactive materials may react violently in contact with air or oxygen or with compounds containing oxygen. However, air contains other chemicals, one of which is water vapor. Nitrogen and the trace gases are largely inert. Air-reactive substances may produce significant heat, fire, or toxic gases. End products of the reaction are likely to be hazardous. Air-reactive substances are pyrophoric, self-heating, or self-reacting.

Pyrophoric materials will burn when exposed to air. Self-heating materials will react when exposed to air without an external energy source. These materials are also referred to as spontaneously combustible. The flame of some pyrophoric materials is clear and not readily visible.

The U.S. Department of Energy (DOE) defines pyrophoric materials as those that immediately burst into flame when exposed to air. DOE defines spontaneously combustible materials as those that will ignite after a slow buildup of self-generated heat.

The U.S. Department of Transportation (DOT) defines a pyrophoric material as a solid or liquid material that will burn without an ignition source within five minutes of exposure to air according to United Nations specifications. DOT defines self-heating materials as those that exceed 200°C or self-ignite within 24 hours under certain test criteria. These specifications are for defining purposes as they relate to the safe transportation of dangerous goods and are not intended for characterizing hazards of unlabeled substances.

Trying to decide if a material is pyrophoric or merely spontaneously combustible becomes a moot point if 10 gallons of the stuff is spilled and already burning. This book emphasizes hazards that can impact you and others. You are more likely interested in predicting the behavior of an unknown substance in order to protect people, facilities, and equipment.

Air-reactive materials are sometimes packaged in gas cylinders, although they may not be gases themselves, or they may be packaged under nitrogen or some other inert atmosphere. In general, air-reactive materials are identified by exposing a small amount of substance to the air and observing. Rapid discoloration is suggestive of reaction with air. Some of these materials may not ignite immediately, but you can detect an increase in temperature with a thermometer, preferably an infrared model. Expect an air-reactive compound to emit toxic fumes or smoke.

Air-Reactive Alkali Metals
Alkali metals such as cesium, rubidium potassium, sodium, and lithium are air reactive and are described in order of decreasing reactivity. Cesium and rubidium react vigorously with air at room temperature to form the metal oxide, which results in a self-sustaining metal fire that rapidly heats up to almost 2,000°C. Potassium is less reactive at room temperature and generally will not ignite spontaneously. Bulk sodium and lithium will not result in a fire because they are even less reactive and oxidation occurs slowly. However, if these less reactive metals are finely divided or heated (the rate of reaction doubles for every 8–11°C [15–20°F] increase in temperature according to the U.S. DOE) and then exposed to air, spontaneous ignition and self-sustaining fire may occur. When these metals burn, the resulting metal oxide immediately condenses to form a dense, white fume that is highly corrosive to the lungs, eyes, and skin. These metal oxides form metal hydroxides that can be detected by placing a small amount in water and testing with a pH test strip. A pH test strip wetted with deionized or distilled water will indicate strong base if placed in the fume emitting from the burning sample.

Under various circumstances, alkali metals if cut or scraped may react to form unstable, higher oxides (e.g., peroxides or superoxides) that may react if disturbed. These higher oxides can react with the base metal or organic materials in an explosive manner or can begin burning. In some cases, they may be shock sensitive. The higher oxide crust may be white, yellow, or orange. If a peroxide or superperoxide is suspected, it could possibly be detected with a wet potassium iodide starch test strip or a wet peroxide test strip. Refer to the section on reactive materials related to peroxide detection.

Sodium oxidizes rapidly in moist air, but spontaneous ignitions have not occurred unless the sodium is finely divided. Sodium ignites at about 880°C (1,616°F), burns vigorously and forms a dense white cloud of caustic sodium oxide fume. A pH strip wetted with deionized or distilled water will indicate strong base if placed in the fume emitting from the burning sample.

Potassium is more reactive than sodium and will react with atmospheric oxygen to form three different oxides: potassium oxide (K_2O), potassium peroxide (K_2O_2), and potassium superoxide (KO_2). These higher oxides can react with the base metal or organic materials and burn intensely, or they could be shock sensitive and explode. Use very small amounts of sample when testing. A pH strip wetted with deionized or distilled water will indicate a strong base if placed in the fume emitting from the burning sample.

NaK (commonly pronounced like the word "knack") is any number of sodium-potassium alloys that are liquids or low melting point solids. The potassium oxides named above can form a crust over the NaK. If the crust is pierced and the potassium oxides mix with the NaK, a high temperature reaction similar to a thermite reaction can occur. This observation is more likely to be experienced if the crust is accidentally pierced and not necessarily through careful manipulation. Test small amounts of crust using a method described above.

Lithium is not pyrophoric, but is listed here as an alkali metal. Lithium ignites and burns at 180°C (356°F) as it melts and flows. A pH strip wetted with deionized or distilled water will indicate strong base if placed in the fume emitting from the burning sample.

Zirconium in bulk form can dissipate high temperature without igniting, but a cloud of 3 micron dust particles can ignite at room temperature. A layer of 6 micron dust particles will ignite at 190°C (374°F). Metal dust fires have ignited when the dust is disturbed through maintenance and cleaning processes. Finely divided metal dusts such as nickel, zinc, and titanium may be air reactive. These metals are more likely to react if disturbed or mixed into the air.

Hafnium in dust form reacts comparably to zirconium. In sponge form, hafnium may spontaneously combust. Calcium, if finely divided, may ignite spontaneously in air, as will barium and strontium. Aluminum in powdered form can ignite spontaneously in air.

Uranium may be pyrophoric if finely divided. The source and products of combustion will be radioactive. Plutonium has pyrophoric characteristics similar to uranium, but plutonium dust ignites more readily. According to lessons learned from the 1969 Rocky Flats Plant fire, spontaneous pyrophoricity of plutonium is unpredictable, even in bulk form. Use a gamma detecting instrument to screen metals for radiation. Any of these metals will leave a basic fume or residue that can be detected with a pH test strip. Some of these metals will burn with a characteristic flame color; some emit intense UV light and eye protection may be necessary.

Air-Reactive Hydrides

Hydrides are air reactive and often begin burning if exposed to the atmosphere. Examples include arsine, barium hydride, diborane, and diisobutyl aluminum hydride.

Arsine (AsH_3 or arsenic hydride) is a gas and will generally not ignite in air unless at elevated temperatures, but it can be detonated by a suitably powerful initiation (heat source, shock wave, electrostatic discharge). Arsine may be carried in other substances. The ignition temperature of many of these arsine containing substances is lower than that of arsine, causing ignition in air even at low temperatures (below 0°C, 32°F). All arsine compounds should be considered pyrophoric until they are properly characterized. Arsine may be detected with a colorimetric air monitoring tube. Some colorimetric tubes are sensitive to both phosphine and arsine.

Diborane (B_2H_6) is a highly reactive gas with a flammable range of 0.9–98% and an ignition temperature of 38–52°C (100–125°F). Diborane will ignite spontaneously in moist air at room temperature. Diborane is normally stored at less than 20°C (68°F) in a well ventilated area segregated from other chemicals. A sample of diborane can be ignited easily in air with a hot wire or other low temperature source. The flame should be green, but the presence of other materials may obscure the color.

Phosphine (PH_3) is a highly toxic, colorless gas. Phosphine may ignite spontaneously at 212°F. Diphosphine (P_2H_4) is also pyrophoric. Both of these materials are detectable with an AP2C® air monitor, which is highly sensitive to phosphorous and sulfur compounds. The response of the monitor should favor the G-agent alarm, although both light columns may illuminate. If using an air monitor, be sure to begin with a dilute fume sample so as to not overload the monitor. Phosphine can also be detected by the use of a colorimetric air monitoring tube. Some colorimetric tubes are sensitive to both phosphine and arsine.

Silane (SiH_4 or silicon tetrahydride) and disilane (Si_2H_6) are gases that might ignite in air. The presence of other hydrides as impurities causes ignition always to occur in air. Nearly pure silane ignites in air, and mixtures of up to 10% silane may not ignite. Hydrogen liberated from its reaction with air often ignites explosively. All silanes should be considered pyrophoric until they are properly characterized. Silanes will leave a residue similar to the soot of a diesel fuel fire with one striking exception: the soot is white, not black (silicon oxides are essentially glass).

Silane Fire at a Department of Energy Facility

A 20' × 30' room with concrete block walls was used for distribution of process gases to clean room areas. Silane cylinders were located in 12-gauge metal gas cabinets. The ventilation system for the cabinets was in the process of being upgraded and the cabinets were protected by automatic sprinklers.

The silane cylinder involved was installed about 30 minutes prior to the incident. Employees in the area heard a loud "pop" from the process gas distribution room. Upon investigation, they found the windows of the cabinet broken, the doors open, and fire coming from the cylinder valve. The sprinkler activated properly and confined the fire to the cylinder head.

The escaping silane was caused by an improper connection of the cylinder to the distribution piping. There was evidence that the connection was cross-threaded, allowing the leakage. The flow of silane could not be shut off because of damage to the cylinder manifold connections. The fire continued to burn for about eight and a half hours, until all silane in the cylinder had been consumed.

Hydrazine (N_2H_4) is a clear, oily liquid resembling water in appearance and possesses a weak, ammonia-like odor. It may be in aqueous or anhydrous solution. Higher concentrations fume in air. The flash point of hydrazine is 38°C (100°F), and the flammable range is 4–100%. Hydrazine is detected by its high pH if even a minute amount of water is present. Colorimetric air monitoring tubes are available for hydrazine, although some do not discern hydrazine from ammonia and amines (all forming basic gas or vapor) because the tube detects basic pH.

Plutonium hydride (PuH_2 or PuH_3) in finely divided form is pyrophoric. Both the hydride and products of combustion will be radioactive. This material should be screened with a gamma detecting instrument. Like the other hydrides, its fume will present a basic pH.

Barium hydride (BaH_2) is pyrophoric in powdered form. Diisobutylaluminum hydride is air sensitive and water reactive. Both of these materials produce a basic fume, detectible with water moistened pH test strips. A colorimetric spot test for barium and aluminum would identify them if identification is necessary. All of these hydrides will react similarly and pose similar hazards with the exception of radioactive plutonium hydride.

Air-Reactive Phosphorous

Phosphorus may be pyrophoric. White (also called yellow) phosphorus is a colorless to yellow, translucent, nonmetallic solid and ignites spontaneously on contact with air at or above 30°C (86°F). Red phosphorus is not considered pyrophoric but ignites easily to form phosphine, a pyrophoric gas described above. A solid that is able to ignite from contact with a warm wire should have the smoke analyzed for phosphine with a colorimetric air monitoring tube or other air monitor that is sensitive to phosphorous and sulfur compounds. Use a dilute fume sample so as to not overload the monitor.

Other reactions may be observed when the sample is exposed to air. Some of these metals may only discolor in air at room temperature, but if finely divided or heated, they may spontaneously ignite.

Others

Other material may appear to be air reactive. Fuming acids are a good example because they form a visible "smoke" that is more apparent in humid air. Fine dusts and fibers can form a visible cloud if disturbed. These materials are not truly air reactive, but they appear to be. Placing a pH strip moistened with deionized or distilled water into the "smoke" from a fuming liquid will identify the pH of the fume. Basic fumes from air-reactive materials are described above. Acid fumes are most likely a concentrated acid or a reactive product that forms acid on contact with water such as thionyl chloride (Figure 2.3) or phosphorous trichloride. Regardless of pH value, corrosive vapor, fume, and smoke are easily characterized.

Water Reactivity

The term "water reactive" has certain definitions based on transportation, manufacturing, or other areas of interest. Placing the unknown substance in

Figure 2.3 Thionyl-chloride-forming acidic fume in moist air.

contact with water can produce many more reactions than listed by any one definition. This section is concerned with the hazardous reactions formed from an unknown substance in water, but many other reactions may occur that help classify the material. Material that fumes when exposed to air may be reacting with moisture in the air and is actually water reactive and not air reactive.

Water-reactive materials may cause secondary hazards:

- Fire
- Gas or vapor
  - Flammable
  - Corrosive
  - Toxic
  - Oxidizing
  - Asphyxiant
- Corrosive

The simplest way to determine water reactivity is to add a small amount of the sample to water and observe. Be concerned with observing any chemical or physical reaction of the unknown substance with water so its subsequent behavior may be predicted if a larger amount makes contact with water. For example, tetrachlorosilane will evolve as a toxic gas and fit the definition of a Department of Transportation water-reactive, toxic inhalation hazard. Concentrated hydrochloric acid will spontaneously emit white fumes; however, as a fuming acid, it does not fit the criteria as a Department of Transportation – Water-Reactive Toxic Inhalation Hazard (DOT WRTIH).

Substances that are water reactive are dangerous, but the by-products of these reactions are often more hazardous than the original substance. Water-reactive substances often turn the water solution strongly acidic or basic or may begin burning spontaneously. The most severe hazard involves those water-reactive substances that produce a toxic gas or vapor.

The U.S. DOT has done extensive research involving toxic inhalation hazards (TIH) resulting from a transportation accident of any type of hazardous material. Within the TIH group are the water-reactive, toxic inhalation hazard (WRTIH) substances. All of these water-reactive chemicals produce a significantly toxic gas or vapor. Of the water-reactive materials researched for transportation, several are also TIH compounds that produce a secondary, sometimes more toxic TIH gas upon reaction with water.

DOT's *2020 Emergency Response Guidebook* lists response guides more than 100 water-reactive materials, defined here as materials that emit a TIH gas upon contact with water. The TIH gases produced by these water-reactive materials are listed in Table 2.2. These water-reactive materials are allowed to be transported. Other water-reactive material exists that is forbidden from

Table 2.2　Toxic Gases Produced by Materials When Spilled in Water

Symbol	Chemical	Immediately Dangerous to Life or Health (IDLH), ppm	CAS Number
Br_2	Bromine	3	7726-95-6
Cl_2	Chlorine	10	7782-50-5
HBr	Hydrogen bromide	30	10035-10-6
HCl	Hydrogen chloride	50	7647-01-0
HCN	Hydrogen cyanide	50	74-90-8
HF	Hydrogen fluoride	30	7664-39-3
HI	Hydrogen iodide	30*	10034-85-2
H_2S	Hydrogen sulfide	100	7783-06-4
NH_3	Ammonia	300	7664-41-7
NO_2	Nitrogen dioxide	20	10102-44-0
PH3	Phosphine	50	7803-51-2
S02	Sulfur dioxide	100	7446-09-5

*No empirical data available for HI. AEGL values set using HBr.
Source: National Advisory Committee for Acute Exposure Guideline Levels for Hazardous Substances.

transport; these materials are likely to be found engineered into fixed sites or possibly used in an illegal manner.

The researchers excluded from their definition water-reactive substances that produce flammable gases that do not otherwise pose a toxic hazard. Examples would include hydrogen gas from wet calcium hydride and acetylene gas from wet calcium carbide.

Alkali metals react vigorously with water to release flammable hydrogen gas and form the corresponding hydroxide. For example, potassium metal in water would produce hydrogen gas and potassium hydroxide. The resulting solution is strongly basic. The rate of reaction increases as atomic weight increases. Therefore, lithium metal reacts more slowly than sodium metal, which reacts more slowly than potassium metal. Considerable heat is generated quickly from these reactions and may be sufficient to ignite the hydrogen gas.

Others

Other reactivity hazards come from materials that can self-react, decompose, or otherwise easily release energy or chemically change. Examples include aluminum powder and matches that can ignite by friction; solidified alcohol (Sterno®) that is easily ignitable; benzene sulphohydrazide that is self-reactive; wetted zirconium powder, a desensitized explosive; and peroxidizable material, which may be shock sensitive. Peroxides are also included in the section on oxidizers.

Material demonstrating reactivity hazards can produce secondary hazards:

- Spontaneous combustion
  - Fire
  - Explosion
  - Toxic gas
- Violent reaction
  - Vigorous reaction
  - Polymerization
  - Decomposition

Some materials form organic peroxides under certain conditions or simply by aging. Other materials can polymerize upon the loss of an inhibitor. A few of these materials are susceptible to both polymerization and peroxide formation. Any of these materials may be self-reacting or cause spontaneous polymerization or other violent reaction such as a fire or explosion. These materials can form polyperoxide chains or cyclic oligoperoxides that are difficult to detect; these types of peroxides are usually not detected by common peroxide test strips. The peroxides can come out of solution and form crystals or a gel in the bottom of the container, which may be visible if a light is shined through the glass container. The crystals can appear as fine, glass-like fibers; a film of crystalline material on the container wall; a web of material within liquid; or as a lump of salt-like material in the bottom of the container. Gel can appear as a lump or thick layer in liquid. They are extremely unstable and can violently decompose with the smallest disturbance, sometimes even spontaneously. Some peroxides will burn violently while others will produce low order explosions and either reaction may be initiated by shock, spark, or heat.

The likelihood of peroxide formation can be estimated based on age of the substance and the amount of evaporation. The scope of this book is limited to unknown material, so it is assumed an intact label is not available. If evaporation is determined to be greater than 10%, significant concentration of peroxide or polymerizing material may be present.

Peroxides can be formed by organic and inorganic compounds. Organic peroxide-forming chemicals may be:

- Ethers and acetals
- Olefins with allylic hydrogen, chloro- and fluoroolefins, terpenes, and tetrahydronaphthalene
- Dienes and vinyl acetylenes
- Aldehydes
- Ureas, amides, and lactams
- Vinyl monomers (vinyl halides, acrylates, methacrylates, and vinyl esters)

Inorganic peroxide-forming chemicals may be:

- Alkali metals (particularly potassium)
- Alkali metal alkoxides and amides
- Organometallics

These types of compounds describe a large number of individual substances. However, there are more common compounds that are more likely to be encountered. There are three groups of common materials listed below that can polymerize, form peroxides, or do both. The most hazardous compounds are the first group, those that form peroxides in storage without being concentrated. These are materials that can accumulate hazardous levels of peroxides simply in storage after exposure to air. Substances in this first group should always be checked for peroxides. Unfortunately, there is no single test for all hazardous peroxides and polymer-forming material.

A concentration of less than 80 ppm peroxide is considered safe. Consult a safety data sheet (SDS) or manufacturer before opening a container if one of these compounds is suspected. These materials will present observable characteristics in contact with water that identify hazards or suggest further simple testing.

These materials form peroxides in storage without being concentrated:

- Sodium amide
- Potassium amide
- Vinylidene chloride
- Isopropyl ether
- Potassium metal
- Butadiene
- Divinyl ether
- Divinyl acetylene

This second group of solvent materials can form soluble hydroperoxides and ketone peroxides that can be a hazard if heated or concentrated. Consult an SDS or manufacturer before opening a container if one of these compounds is suspected:

- Furan
- Isopropanol
- Methylcyclohexane
- Methyl isobutyl ketone
- Tetrahydrofuran
- Tetrahydronaphthalene
- Dicyclopentadiene
- Diglyme

- Diethyl ether (ethyl ether)
- 1,4-Dioxane
- Ethylene glycol dimethyl
- Ether (monoglyme)
- Acetal
- 2-Butanol
- Cellosolves (e.g., 2-ethoxyethanol)
- Cumene (isopropylbenzene)
- Cyclohexene
- Decalin (decahydronaphthalene)

This third group of materials includes liquids and liquefied gases that can be initiated by oxygen to polymerize. These materials are safe to use or store if the peroxide concentration is less than 80 ppm. Consult an SDS or manufacturer before opening a container if one of these compounds is suspected:

- Acrolein
- Acrylic acid
- Acrylonitrile
- Chloroprene (2-chlorobutadiene)
- Chlorotrifluoroethylene
- Ethyl acrylate
- Ethyl vinyl ether
- Styrene
- Tetrafluoroethylene
- Vinyl acetate

Corrosives

Corrosivity hazards are presented by materials that exhibit high corrosion rates on metal or human tissue (DOT) or have an extreme pH. DOT Class 8 corrosive materials are the second most commonly shipped hazardous material (after flammable liquids and gases) in the United States.

Corrosive hazards can produce secondary hazards:

- Fire
- Toxic gas or vapor
- Burns
  - Skin and eye
  - Respiratory

Corrosives are used in several areas of commerce, and not all are transportable. Transportation data are nonetheless valuable in predicting the relative volume of hazardous material in use. Table 2.3 indicates shipments of hazardous material by class and by inference the likelihood of encountering an unknown sample.

Corrosive material that presents a fire hazard may be a combustible material or it may ignite combustible material through reaction. All organic acids are fire hazards. Amines are organic bases and are also fire hazards. Additionally, many concentrated inorganic acids may be able to ignite ordinary combustible material by chemical reaction and dehydration. A term called "heat of hydration" describes the heat released by a compound as it absorbs water (Figure 2.4). Some compounds have a heat of hydration high enough to boil the water it contacts. This explains the violent spattering caused when a small amount of water is added to a concentrated acid. Strong bases might, but are not as likely, to cause a fire due to reaction. Major characteristics of some common acids and bases are shown in Table 2.4.

Table 2.3   U.S. Hazardous Materials Shipments by Hazard Class, 2012

Hazard Class and Description	Value		Tons		Ton-Miles
	Billion $	Percent	Millions	Percent	Billions
Class 1. Explosives	18	0.8	4	0.2	1
Class 2. Gases	125	5.4	165	6.4	33
Class 3. Flammable liquids	2,017	86.4	2,203	85.4	205
Class 4. Flammable solids	5	0.2	11	0.4	6
Class 5. Oxidizers and organic peroxides	8	0.3	12	0.5	5
Class 6. Toxics (poison)	15	0.7	8	0.3	4
Class 7. Radioactive materials	12	0.5	S	U	0
Class 8. Corrosive materials	76	3.2	125	4.9	38
Class 9. Miscellaneous dangerous goods	58	2.5	51	2	16
Total	2,334	100	2,580	100	308

KEY: U = data are not available or less than one unit of measure or rounds to zero; S = data were not published because of high sampling variability or other reasons.
Note: Numbers may not add to totals due to rounding.
Source: U.S. Department of Transportation, Bureau of Transportation Statistics, and U.S. Department of Commerce, Census Bureau, 2012 Commodity Flow Survey, American Fact Finder, Hazardous Materials (Washington, DC: December 2014).

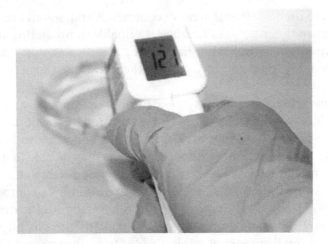

Figure 2.4 Infrared thermometer measuring increase in temperature caused by addition of sodium hydroxide to water.

Table 2.4   Major Characteristics of Some Common Acids and Bases

Material	Characteristics
Concentrated sulfuric acid	Aggressively destroys organic material, high production of heat when mixed with water
All organic acids	Volatile, fire hazard
All organic bases	Volatile, fire hazard
Nitric acid	Strongly oxidizing, volatile, may be explosive, high production of heat when mixed with water
Perchloric acid	Strongly oxidizing, explosive if dehydrated, volatile
Hydrochloric acid	Dissolves most metals readily, volatile
Hydrofluoric	Volatile, highly reactive
Sodium hydroxide	High production of heat when mixed with water, destructive to organic material
Potassium hydroxide	Similar to sodium hydroxide
Ammonium hydroxide	Volatile, destructive to organic material

Corrosivity hazards are determined mainly by pH value. Water must be present to detect pH with common test methods. Since the human body is about 80% water, it is no surprise that solid or liquid corrosive material will cause burns. Corrosive and caustic solids, liquids, and gases or vapors can cause significant injury through contact or inhalation.

Oxidation

Oxidation hazards come from materials that can cause another material to burn or explode. Oxidizers can also generate gas quickly.

Oxidizers can produce secondary hazards:

- Fire
- Gases or vapors
  - Oxidizing
  - Corrosive
  - Toxic
  - Explosive

The most familiar form of oxidation is fire, but oxidation has a broader context in terms of chemical reactions. Since the term "oxidation" is relative to each situation, it becomes difficult to define which compounds are always an oxidation hazard. When assessing an oxidation hazard, it must always be determined within the context of another material that may react with it.

Oxidation occurs when a substance loses one or more electrons. The opposite, reduction, occurs when a substance gains one or more electrons. Collectively, this is referred to as a "reduction oxidation," or "redox" reaction. In a redox reaction, one substance loses electrons and the other substance gains electrons. In other words, one is oxidized and the other is reduced.

For example, substance A and B participate in a redox reaction. A loses an electron and becomes A^+. B gains the electron and becomes B^-. A has been oxidized and B has been reduced. That means A is the reducing agent and B is the oxidizing reagent.

Redox reactions occur in a wide venue, which includes fire, biochemical processes, acid base neutralization, batteries, metal fluxing, etc. Just because it is a redox reaction doesn't mean it will burn. Since we are organic beings, some oxidizers that are not able to spontaneously burn would be able to react with body tissue and cause significant injury.

Hazardous oxidation reactions that occur in the field often are the result of an oxidizing substance contacting a fuel or a reducing agent. Most people can name a long list of powerful fuels (reducers), such as gasoline, propane, acetylene, paper, rubber, etc. When asked to name a list of powerful oxidizers, the short list seems to stop at oxygen. Other oxidizers, while not as familiar, certainly exist, such as hydrogen peroxide, nitric acid, chlorine, bleach, perchlorate compounds, etc. Some substances can form peroxides under certain conditions; these are oxidizing hazards, but are included in the section on reactivity. Since fuels are so common in our environment, it only makes sense that the more powerful an oxidizer is, the greater the safeguards to prevent it from reacting spontaneously.

Case Study: Nitrate Induced Acrolein Fire

An acrolein synthesis plant in Niihama, Ehime, Japan experienced a severe fire on December 23, 1998. A nitrate fluid was used as a coolant for production of acrolein. In a closed system, the nitrate was able to effectively cool the reactor during high-temperature operation. Work in the plant had reduced the coolant flow rate to about half and temperature of the coolant rose to an estimated 500°C. The heat exchanger failed and the acrolein and nitrate mixed through the opening. The resulting high-temperature reduction oxidation reaction overpressurized the system and blew open a rupture disk. The reactor exploded and the ensuing fire destroyed the factory. The plant was located at the base of a steep, wooded hill that was ignited by venting material. The ensuing fire destroyed over 1000 m² of a wooded hillside. It took 126 firefighters with 26 vehicles to extinguish both fires in about four hours.

Below is a partial list of oxidizers and relative oxidizing potential.

Increase Rate of Combustion

- Aluminum nitrate
- Perchloric acid 60% or less
- Ammonium persulfate
- Potassium chlorate
- Barium chlorate
- Potassium dichromate
- Barium peroxide
- Potassium nitrate
- Calcium chlorate
- Potassium persulfate
- Calcium nitrate
- Silver nitrate
- Calcium peroxide
- Silver nitrite
- Cupric nitrate
- Sodium perborate
- Hydrogen peroxide

- Sodium perchlorate
- Lead nitrate
- Sodium persulfate
- Lithium hypochlorite
- Strontium chlorate
- Lithium peroxide
- Strontium nitrate
- Magnesium nitrate
- Strontium nitrite
- Magnesium perchlorate
- Thorium nitrite
- Magnesium peroxide
- Uranium nitrate
- Nickel nitrate
- Zinc chlorate
- Nitric acid 70% or less
- Zinc peroxide

Cause Spontaneous Ignition
- Calcium hypochlorite
- Sodium chlorite (> 40%)
- Chromic acid
- Sodium peroxide
- Hydrogen peroxide (27.5–52%)

- Sodium permanganate
- Nitric acid
- Trichloroisocyanuric acid
- Potassium bromate
- Sodium dichloroisocyanurate
- Potassium permanganate

Decompose with Catalyst or Heat
- Ammonium dichromate
- Perchloric acid (60–72.5%)
- Hydrogen peroxide (52–91%)
- Potassium dichloroisocyanurate
- Calcium hypochlorite (>50%)
- Sodium dichloroisocyanurate

Cause Explosive Reaction When Exposed to Catalyst, Heat, Shock, or Friction
- Ammonium perchlorate
- Perchloric acid
- Ammonium permanganate
- Potassium superoxide

Oxidizers are affected not only by a reducer, but also by heat, pressure, and concentration. An example is perchloric acid, which in a concentration of 60% or less will cause accelerated burning of ordinary combustibles, such as wood, paper, and cloth. Perchloric acid (60–72.5%) will decompose with heat (without a reducer) and release thermal decomposition products, mostly oxygen. Perchloric acid that has crystallized has been documented to explode due to shock or friction. Other incompatible materials, such as contaminants that act as a catalyst, can cause explosive decomposition.

Explosion

Explosive material is hazardous due to its ability to physically disrupt a person's body by producing a shock wave. Explosives can present secondary hazards:

- Fragmentation
- Toxicity
  - Nitrates
  - Heavy metals
  - Production residue
- Fire

When the word "explosive" is used as a noun, it suggests an object that meets a certain set of commercial intentions. For example, the U.S. DOT defines an explosive as "any chemical compound, mixture, or device, the primary or common purpose of which is to function by explosion, i.e., with substantially instantaneous release of gas and heat" (29 CFR 1910.109(a)(3)) followed by exceptions and examples.

The technician attempting to determine the explosive hazard of an unknown substance is much more interested in using the word "explosive" as an adjective – that is, the violent bursting out of heat, gas, light, sound, and fragments.

An unknown substance may explode by rapid combustion, detonation, or some physical process that produces a sudden production of gas beyond the strength of the container. In any case, the technician is deeply concerned with identifying the degree of explosive potential of an unknown substance in a particular situation.

Rapid combustion is characteristic of manufactured low explosives, such as black (gun) powder. Many other substances that are not defined as explosives have explosive potential such as smokeless gunpowder, a plume of hydrogen gas in air, and a house full of natural gas. The significant thing that makes the situation explosive is the rate of combustion in conjunction with confinement of the substance. Whether the unknown substance reacts in the confines of a steel gun barrel, a wood framed house, or the light mass of the atmosphere, the explosive hazard is significant in each case.

Detonation does not require confinement of the substance in order to be explosive. A detonation shock wave moves powerfully through the explosive material and provides force beyond that which might be provided by a container.

Some chemical reactions produce gas that can increase pressure within a container. If the pressure exceeds the strength of the container, an explosion can occur if the container fails catastrophically. Many polymerization reactions can produce gas by design, such as closed cell foam applications. Others simply produce gas from the addition of certain reagents or a change in pH, such as acid added to a bicarbonate compound. In any case, consider a pressure release explosion when choosing a container to hold a sample.

Volatile liquids (and to a lesser extent nonvolatile liquids and solids) may be heated rapidly and change phase to a gas. Many liquids have defined expansion ratios that predict the volume of gas produced by a liquid converted to a gas. For example, one gallon of water will boil away to form about 1,700 gallons of steam. Cryogenic nitrogen has an expansion ratio of 1:691, and carbon dioxide expands from liquid to gas at a ratio of 1:845. Liquefied petroleum gas (LPG) has an expansion ratio of 1:270. In certain scenarios without an ignition source, water may have the highest explosion potential.

Table 2.5   Some Mixtures of Fuels and Oxidizers Intended to Be Explosive

Explosive	Mixture
Black powder	Sulfur, charcoal, potassium nitrate
Flash powder	Fine metal powder, usually aluminum or magnesium, and a strong oxidizer such as potassium chlorate or perchlorate
ANFO	Ammonium nitrate and fuel oil
LOX mixtures	Combustible material and liquid oxygen (LOX)
Armstrong's mixture	Potassium chlorate and red phosphorous

A boiling liquid, expanding vapor explosion (BLEVE) is a hazard pos-sible with nearly all liquids in intact containers without adequate pressure relief equipment. A BLEVE might occur in a field analysis setting if a tightly sealed sample container holds a self-reactive liquid that warms above its maximum safe storage temperature (MSST) and self-degrades. Substances that spontaneously degrade at room temperature must be stored below room temperature. This observation at the site of discovery of the unknown sub-stance should prompt the technician to keep the sample cooled.

Explosive substances may be homogenous compounds, such as manu-factured explosives, or they may be mixtures of fuel and oxidizer (Table 2.5).

Examples of homogenous explosive compounds include:

- Nitroglycerine and dynamite
- Triacetone triperoxide (TATP)
- Trinitrotoluene (TNT)
- Hexahydro-1,3,5-trinitro-s-triazine (RDX)
- Pentaerythritol tetranitrate (PETN)
- Nitrocellulose

DOT Explosive Classes

"Class A explosives." Possessing, detonating, or otherwise maximum hazard; such as dynamite, nitroglycerin, picric acid, lead azide, fulminate of mercury, black powder, blasting caps, and detonating primers. 1910.109(a)(3)(i)

"Class B explosives." Possessing flammable hazard, such as propellant explosives (including some smokeless propellants), photographic flash powders, and some special fireworks. 1910.109(a)(3)(ii)

"Class C explosives." Includes certain types of manufactured articles that contain Class A or Class B explosives, or both, as components but in restricted quantities. 1910.109(a)(3)(iii)

"Forbidden or not acceptable explosives." Explosives that are forbid-den or not acceptable for transportation by common carriers by rail freight, rail express, highway, or water in accordance with the regulations of the U.S. Department of Transportation, 49 CFR chapter I. 1910.109(a)(3)(iv)

"Water gels or slurry explosives." These comprise a wide variety of materials used for blasting. They all contain substantial proportions of water and high proportions of ammonium nitrate, some of which are in solution in the water. Two broad classes of water gels are (i) those that are sensitized by a material classed as an explosive, such as TNT or smokeless powder, (ii) those that contain no ingredient classified as an explosive; these are sensitized with metals such as aluminum or with other fuels. Water gels may be premixed at an explosives plant or mixed at the site immediately before delivery into the borehole. 1910.109(a)(18)

"Pyrotechnics." Any combustible or explosive compositions or manufactured articles designed and prepared for the purpose of producing audible or visible effects which are commonly referred to as fireworks. 1910.109(a)(10)

Explosive Composition

Table 4.5 in Chapter 4 lists characteristic composition of explosives. You can use the table to rule in or out possible identities of suspected explosive material. Chemical traits or functional groups can be detected by several simple tests. Table 2.6 describes explosive characteristics detected by these tests.

Table 2.6   Characteristics of Explosive Material Revealed by Qualitative Tests

Analysis Step	Characteristic
Ignition as described for flammable solids	Hydrocarbon
Ignition as described for flammable solids	Nitro group
Ignition as described for flammable solids	Organic nitrate
Nitrate test strip	Nitrate anion
Methylene blue	Perchlorate
Peroxide test strip, potassium iodide starch test strip	Peroxide
Heavier than water	Metal
Oxygen sensor in a multisensor detector	Oxygen
Temperature	LOX (liquid oxygen)
Ignition as described for flammable solids	Picrate
Ignition as described for flammable solids	Unstable structure

Fire

Fire hazards are more familiar to most people than oxidizer hazards. Fire hazards generally exist in the context of the fuel acting as a reducing agent and atmospheric oxygen acting as the oxidizer. When characterizing an unknown material as a fire hazard, determine how readily it ignites and what it produces as combustion by-products.

Fire hazards may present secondary hazards:

- Fire spread
- Combustion products
  - Corrosive
  - Toxic

Types of fuel include spontaneously combustible, flammable, combustible, and ordinary combustible materials. All of these materials will burn, but what sets them apart is the temperature at which they may be ignited. They are often similar in that they nearly always contain organic material. Hydrocarbons will burn with air to produce carbon dioxide and carbon monoxide. In the case of less efficient combustion, volatized fuel and partially combusted fuel may be present.

Fuels that contain elements other than hydrogen and carbon produce an array of by-products. Depending on several variables, such as reducer concentration, oxidizer concentration, temperature, etc., by-products of various combinations and toxicity may be present.

Halogenated hydrocarbons containing fluorine, chlorine (most commonly), bromine, or iodine produce hazardous by-products when subjected to fire conditions. Halogens in general have higher electronegativity than oxygen and resist yielding to the oxidative effect of air; but high heat can force the halogen away from the hydrocarbon. The result can produce unburned vapor, halogenated fragments, and halogen gas or acid. The halogen acids in higher concentration can be detected with a wet pH test strip in the smoke.

Fuel containing nitrogen can produce a variety of oxidized nitrogen compounds that may have various combinations with oxygen. These compounds are often referred to as NOX or NO_x, where x is the variable number of oxygen atoms that may combine with nitrogen as combustion by-products. They may be detectible in smoke with a nitrate test strip.

Sulfur, phosphorus, and other nonmetals present in organic substances can form corresponding oxides that may form toxic and corrosive compounds in the presence of water. These may be present as acid-producing gases detectable by pH test strip in smoke or as various oxides, some of which may be detectable by colorimetric air monitoring tube.

Fuel that is liquid or gas can cause fire spread and pose a mobility hazard. Mobile gas or liquid may travel a considerable distance before encountering an ignition source and releasing energy through fire.

Flammable Gases

When characterizing an unknown gas as a fire hazard, determine how readily it ignites and the characteristics of combustion products.

Flammable gases are in gas phase at standard temperature and pressure. They may be stored as liquefied gases or pressurized gases. When released to air from a pressurized container, gases will be cooler than the product in the cylinder. As a result, the cool gas will have a higher density than that stated in resources, which base data on standard temperature and pressure.

Most gases are heavier than air. Only a few gases have lighter-than-air density at standard temperature and pressure (see Table 2.7).

Density of a gas or vapor is calculated by dividing the mass of the gas by the mass of air, which is about 29. The result is a simple ratio that might be used to predict the buoyancy of the gas or vapor in air. The measurement of vapor density in reference material is accurate only under the conditions specified in the reference. Often the data are based on pure product – that is, 100% concentration of product compared to 100% air.

Table 2.7   Gases That Are Lighter Than Air

Compound	Mass (air = 29)	Vapor Density [100% conc.] (air = 1)
Hydrogen	2	0.07
Helium	4	0.14
Methane	16	0.55
Ammonia	17	0.59
Hydrogen fluoride	20	0.69
Neon	20.2	0.70
Methyllithium	21.9	0.76
Acetylene	26	0.90
Hydrogen cyanide	27	0.93
Diborane	27.6	0.95
Nitrogen	28	0.97
Carbon monoxide	28	0.97
Ethylene	28	0.97

The density of air specified in terms of the reference is dry, pure air at 1 atm and 68°F. The density of the hazardous gas or vapor is also assumed to be pure at 1 atm and 68°F. When variations are made to these specified conditions, the vapor density value changes.

During the uncontrolled release of hazardous material, emergency responders are not interested in predicting the behavior of a single molecule or even a small amount of material. The behavior of an entire plume is important to public safety. Understanding the variables will help you more accurately predict events.

Humidity can affect the density of air. Water vapor will dilute the concentration of pure air. Since clouds exist in the sky, you could guess that water vapor is lighter than air. Water has a vapor density of 0.62 (18/29) and is lighter than pure air, so humid air would be lighter than pure air. Therefore, humid air would tend to allow hazardous gases and vapors to be relatively heavier than when measured in pure air.

Dilution of the hazardous gas or vapor with ambient air will move the vapor density closer to that of air. For example, imagine two 1 liter balloons each filled with a different gas. One balloon is filled with pure, dry air and has a vapor density of 1. The other balloon is filled with hydrogen and has a vapor density of 0.07. If released, the hydrogen balloon would rise rapidly due to the large difference in densities. But what if the hydrogen balloon contained 25% hydrogen and 75% air? To calculate the vapor density in the balloon, the entire sample must be considered:

$$(25\% \times 0.07) + (75\% \times 1) = 0.018 + 0.75 = 0.768$$

The sample in the balloon containing 25% hydrogen has a vapor density of 0.768 and will rise much more slowly than the balloon containing a sample of 100% hydrogen. The more dilute a hazardous material, the more air like it becomes.

Temperature will affect the density of gas or vapor compared to air. Smoke from a campfire rises rapidly due to the extreme temperature difference between the products of combustion and ambient air. A hot plume of hazardous gas or vapor can likewise punch a pathway through the air, rise, and become more air like as it assumes ambient temperature. A heavier-than-air gas released under fire conditions will rise in the plume, move downwind, and fall out of the plume as it assumes ambient temperature. A lighter-than-air gas released from a high-pressure container will sink, then warm and rise to dissipate in the atmosphere. Air currents from wind or ventilation devices can force gases and vapors into other areas.

Flammable gases are easily detected with a combustible gas indicator (CGI). A CGI is commonly incorporated into multiple sensor monitors. This sensor is dependent on normal atmospheric oxygen concentration to function accurately. If oxygen concentration is lower than normal, CGI readings will be depressed. CGI readings are increased if oxidizer concentration in air is high. When testing a flammable gas sample that has been drawn from an enclosed space through tubing, consider the CGI display in the context of the monitor oxygen concentration display to determine an accurate reading. Another method would involve pointing a very low flow of gas sample through atmospheric air toward the CGI sensor; the CGI reading will not be quantitatively accurate, but the presence of a flammable gas in a very low oxygen atmosphere will be identified.

Flammable Liquids

When characterizing an unknown liquid as a fire hazard, determine how readily it ignites and the characteristics of combustion products.

Certain regulations have placed more than one definition for the terms "flammable" and "combustible" based on flash point and other characteristics. In this section, the terms "flammable" and "combustible" simply mean the substance can burn and should not be construed to mean material assigned to a certain shipping class, etc.

Flammable liquids are typically hydrocarbons. Silanes are the only exception likely to be encountered. Flammable liquids composed of hydrocarbons with functional groups may or may not be soluble in water depending on functional groups present in relation to the hydrocarbon structure.

Burnable liquids that are not soluble in water are hydrocarbons such as hexane, gasoline, kerosene, diesel fuel, paint thinner, etc. Molecules containing only carbon and hydrogen are only minutely soluble in water, and this solubility is indistinguishable in field testing and observation. Functional groups may be added to hydrocarbons and the flammability as well as solubility may be affected (Table 2.8). The effect of the functional group on the material properties compared to the hydrocarbon structure is proportional. For example, a single hydroxyl group on a 20-carbon chain will not significantly affect water solubility.

Silanes are the silicon-based equivalents of hydrocarbons. Silicon is located just below carbon in the Periodic Table.

Flammable liquid vapors can be detected in a method similar to that described for flammable gases. Much more information can be gathered if the liquid is actually burned. Observing the ignition and subsequent flame and smoke characteristics will provide significant information about the flammability hazard of an unknown chemical. Additionally, many clues can be discovered that indicate subclasses of flammable liquids.

Table 2.8 Organic Family Relative Effect on Combustion and Water Solubility

Family	Combustion Effect	Water Solubility Effect
Nitro	Increase	Slight increase
Alcohol	Increase	Increase
Nitrile	Decrease	Slight increase
Amine	Decrease	Increase
Ketone	Increase	Increase
Aldehyde	Increase	Increase
Ether	Increase	Slightly increase
Aromatic	Decrease	Decrease
Halogen	Decrease	Decrease

Flammable Solids

Like flammable liquids, flammable solids are often composed of organic material. Other solid material that will burn, i.e., inorganic, includes metals, sulfur, and other material.

Flammable solids are defined for transportation purposes by DOT in three main groups:

- Explosives desensitized by water, alcohol, plasticizers, etc.
- Thermally unstable self-reactive material that can strongly react even without oxygen
- Readily combustible solids

Certain regulations have placed more than one definition for the terms "flammable" and "combustible" based on flash point and other characteristics. In this section, the terms "flammable" and "combustible" simply mean the substance can burn and should not be construed to mean material assigned to a certain shipping class, etc. When characterizing an unknown solid as a fire hazard, determine how readily it ignites and the characteristics of combustion products.

Flammable solids are typically hydrocarbons, but flammable solids include metal powders, sulfur, and other solids that can burn. Flammable solids composed of hydrocarbons with functional groups may or may not be soluble in water, depending on functional groups present in relation to the hydrocarbon structure.

Molecules of combustible solids containing only carbon and hydrogen are only minutely soluble in water and this solubility is indistinguishable in field testing and observation. Functional groups may be added to hydrocarbons and the flammability as well as solubility may be affected. The effect of the functional group compared to the hydrocarbon structure is proportional. For example, a single hydroxyl group on a 20-carbon chain will not significantly affect water solubility. A simple sugar, such as glucose, contains a much higher proportion of hydroxyl to carbon and is a combustible solid that is highly water soluble.

Flammable solids may ignite spontaneously or with a simple spark or other ignition source. Some solids need a little more robust input of energy to burn. Any material that can be forced to burn will reveal characteristics of its structure.

Toxins

Toxic Industrial Material

Toxicity hazard must be considered in terms of not only the substance, but the dose. Humans can receive a dose of toxin through any combination of four routes:

- Inhalation
- Absorption (skin and eye)
- Ingestion
- Injection

The most effective dosing route overall is inhalation. The respiratory tract contains a proportionally higher surface area than that represented by the other routes. Additionally, respiratory tissue is thin and absorptive. This route is protected by the proper fit and use of respiratory protection, such as self-contained breathing apparatus. Skin and eye contact are the most likely route for most activities involving hazardous material if the material can permeate, penetrate, or degrade personal protective equipment (PPE), or if the PPE is improperly used or fitted. A person may or may not have signs and symptoms of a toxic material being absorbed through this route. Ingestion is not likely unless it is inadvertent, for example, smoking or eating without washing or swallowing respiratory mucous after improper respiratory protection. Injection is most likely from accidental puncture by a contaminated sharp object or by fragmentation.

Many toxicity hazards are identified through previous hazard characterization tests. It is safe to assume any material that is evaporating from an unknown release should not be inhaled. Inhalation of smoke produced from a material found to be combustible should be avoided. Materials producing corrosive vapors or fumes should not be contacted in any manner. Water and air-reactive material is certainly toxic, as are reaction products. Some specific gases or vapors may be identified in previous steps. In any case, there is no test for specific toxicity that can be done in the field, but toxicity can be subjectively characterized or, if the substance is identified exactly, toxicity studies can be referenced.

Case Study: Accidental Cyanide Poisoning

An artist in New York had been using a sodium cyanide solution while working with metal. Later, when investigating an odor believed to be coming from the basement, he accidentally knocked over the five-gallon bucket of the solution. A family member called 911 and firefighters found the pulseless artist. He was resuscitated, but never regained consciousness and died after several days in a coma. The hazmat team estimated airborne hydrogen cyanide at 200 ppm, four times the IDLH. Over 100 pounds of sodium cyanide were removed from the basement.

Chemical Warfare Agents

Chemical warfare agents, as defined previously, consist of a relatively small but extremely toxic list of chemicals. Chemical warfare agents present two main hazards: incapacitation of the nervous system or extremely severe destruction of tissue from a very small amount of agent. These agents may present other hazards such as corrosive residue or a flammable gas produced upon decomposition; these hazards are slight relative to the extreme toxicity of chemical warfare agents.

Toxic material used for the purpose of terrorism may or may not include chemical warfare agents. Material that is not on the Schedule 1 List from the Chemical Weapons Convention is generally less toxic by volume, but when considered in terms of lethality, a greater dose is necessary to achieve the lethal effect. Material such as chlorine gas or hydrogen cyanide gas could be used for the purpose of terrorism and is just as deadly as banned chemical warfare agents as long as a lethal dose is achieved.

Biological Warfare Agents and Pathogens

Infectious agents are living creatures and may cause infection in a host organism.

Infection is dependant on a number of factors. The host must be susceptible to infection and susceptibility may be influenced by general health, condition of the immune system, previous infection by similar agents, vaccination, concurrent infection, age, sex, race, etc. The infectious agent is subject to factors such as its ability to survive outside the host, reproductive rate, amount and rate of infusion into the host, etc.

If the host is unable to overwhelm and destroy the invading agent, sickness or death may occur. The infectious agent is contagious if it can be transmitted from one person to another.

A viron (a single virus particle) may infuse genetic material into a host cell, hijack the cell's production facility, and form more virus or a toxic substance. As the virus replicates, it can spread to other areas of the body and rob the cells of essential energy and nutrients. Most viruses are not lethal. Variola major (smallpox) and alphaviruses (Venezuelan equine encephalitis) are examples of potentially lethal viruses.

Bacteria are single-celled organisms that lodge in warm, moist areas within the body and reproduce. Bacterial cells carry their own ability to replicate cells or toxins and are not necessarily dependent on entering the host cells in order to inflict injury. Toxins may cause local or systemic injury. Most bacteria are not lethal. Examples of deadly bacteria include *Bacillus anthracis* (anthrax), *Vibrio cholerae* (cholera), and *Yersinia pestis* (plague).

Rickettsia are a dependent type of bacteria that live in the cytoplasm or nucleus of a host cell. Rickettsia are dependent on the host cell's enzymes for replication. *R. prowazekii* (epidemic typhus), *C. burnetii* (Q fever), and *Rickettsia rickettsii* (Rocky Mountain spotted fever) are rickettsia examples.

Biotoxins are specific and effective poisons produced by bacteria, fungi, algae, plants, and others. These toxins do not replicate and are not alive, but they are produced by living cells. Some toxins may be chemically produced. Examples of biotoxins include botulinum toxin (botulism), tetanus toxin, and ricin.

Certain biological material is common to all biological agents, pathogens, and biological toxins. One of these is protein. An unknown substance that does not contain protein cannot be a biological hazard. Therefore, a protein test can screen unknown substances. Other markers exist, such as adenosine diphosphate (ADP), DNA, and others. Screening tests are useful if the concentration of the target substance is above the detection limit of the test. Also, interference can occur from naturally occurring biological material, so a very low detection limit may produce false-positives.

References

2020 Emergency Response Guidebook, United States Department of Transportation, Washington, DC, 2005.

29 CFR 1910.109, United States Department of Transportation, Code of Federal Regulations, Washington, DC.

49 CFR 173.124, United States Department of Transportation Code of Federal Regulations, Washington, DC.

US Department of Energy, *Accident Case Studies*, Washington, DC, http://www.hss.energy.gov/NuclearSafety/techstds/standard/hdbk1081/hbk1081f.html, accessed February 1, 2007.

Armour, M., Browne, L., and Weir, G., *Hazardous Laboratory Chemicals Disposal Guide*, CRC Press, Boca Raton, FL, 1991.

Bender, H., *Standard Oxidation Potentials*, Clackamas Community College, Oregon City, OR, http://dl.clackamas.cc.or.us/ch105-09/standard.htm, accessed December 23, 2006.

Brown, D.F., Freeman, W.A., Carhart, R.A., and Krumpolc M., *Development of the Table of Initial Isolation and Protective Action Distances for the 2004 Emergency Response Guidebook*, ANL/DIS-05-2, Argonne National Laboratory, U.S. DOE, Argonne, IL, May 2005.

Dominique, M., AIR LIQUIDE S.A., Paris, 2006, http://encyclopedia.airliquide.com.

Ellison, D.H., *Handbook of Chemical and Biological Warfare Agents*, CRC Press, Boca Raton, FL, 1999.

Everett, K., and Graf, F.A., Jr., *Handling Perchloric Acid and Perchlorates, CRC Handbook of Laboratory Safety*, 2nd ed., Steere, N.V, Ed., CRC Press, Boca Raton, FL, January 1971.

Failure Knowledge Database, Case Study, Japan Science and Technology Agency, Saitama, Japan, 1999, http://shippai.jst.go.jp/en/Detail?fn=0&Id=CC120011008&, accessed February 7, 2007.

Houghton, Rick, *Field Confirmation Testing for Suspicious Substances*, CRC Press, Boca Raton, FL, 2009.

How Can You Detect Radiation? Health Physics Society, 2002, http://hps.org/publicinfor-mation/ate/faqs/radiationdetection.html, accessed August 29, 2005.

Laboratory Safety Guide, Understanding Chemical Hazards, University of Wisconsin–Madison Safety Department, Madison, WI, 2006.

Nathan, L., Artist Dies in Basement Cyanide Accident, *Art Hazard News* 14: 1 (1991).

National Institute of Occupational Safety and Health, *NIOSH Pocket Guide to Chemical Hazards*, U.S. Department of Health and Human Services, Centers for Disease Control and Prevention, Atlanta, GA, September 2005.

National Institute for Occupational Safety and Health, *NIOSH Pocket Guide to Chemical Hazards, 3rd Printing*. Centers for Disease Control and Prevention, Washington, DC, September 2007.

Ohio State University Office of Environmental Health and Safety Chemical Hygiene Plan, Table 12 *Partial List of Oxidizers*, Ohio State University, Columbus, OH, http://cfaes.osu.edu/facultystaff/healthsafety/documents/Table12-11-01.pdf, accessed December 22, 2006.

Peroxide Forming Chemicals Management and Assessment Guidelines, University of Washington, Environmental Health and Safety, Seattle, WA, 2005.

Primer on Spontaneous Heating and Pyrophoricity, United States Department of Energy, Washington, DC, December 1994.

Safe Handling of Alkali Metals and Their Reactive Compounds, Environmental Safety and Health Manual, Document 14.7, Lawrence Livermore National Laboratory, Livermore, CA, revised October 13, 2005.

Table 1-63: U.S. Hazardous Materials Shipments by Hazard Class, 2012, U.S. Department of Transportation, Bureau of Transportation Statistics, and U.S. Department of Commerce, Census Bureau, 2012 Commodity Flow Survey, American Fact Finder, *Journal of Hazardous Materials* (Washington, DC: December 2014), http://www.census.gov/svsd/www/cfsmain.html as of December 9, 2014, accessed January 29, 2020.

U.S. Environmental Protection Agency, *Interim Acute Exposure Guideline Levels (AEGLs) for Hydrogen Bromide (CAS Reg. No. 10035-10-6) and Hydrogen Iodide (CAS Reg. No. 10034-85-2)*, National Advisory Committee for Acute Exposure Guideline Levels for Hazardous Substances, Washington, DC, April 2010.

Welcher, F.J. *Organic Analytical Reagents*, D. van Nostrand, New York, NY, 1948.

Detection Technology

3

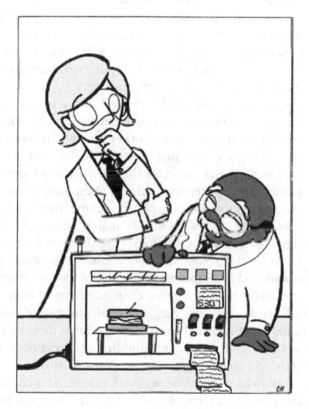

Figure 3.1 Your competency is the most important aspect of qualitative analysis. Use more than one technology to increase confidence and remember to consider the obvious. You don't need an expensive, complicated instrument to identify a chicken sandwich. Be responsive to all possibilities and understand the abilities and limitations of the tests you use. Don't let the instrument make your decisions for you.

Detection Technology

Detection technology is an exciting subject. For hundreds of years, people have sought out tests that could confirm the presence or exclusion of a substance (Figure 3.1). An accurate field test for gold saved a lot of prospectors from wasting their time on fool's gold in the late 1800s. A simple Geiger counter was the technology necessary to discover radioactive material that fueled the atomic age. And who would have thought the U.S. Postal Service would need a test for "white powder?"

Technology available for use by present-day detectors is expanding rapidly from its simple beginnings. As individual detectors are designed for specific purposes, designers often bring more than one technology to bear in an individual detector. Several technological approaches may be used to prepare a sample, manipulate it in some way, analyze it, receive a signal, process the signal, provide a result, and sometimes even process the result for the operator. For example, mass spectroscopy is a fundamental analysis technology used in both field and laboratory analysis, but without a method of separating mixtures, the results are nonspecific. Coupling mass spectroscopy with a gas chromatograph greatly improves performance.

Some equipment is designed to detect a substance in any mixture, such as a hand-held rapid immunoassay that will detect only one specific antigen and will not alert you to the presence of other possibly hazardous biological material. Other equipment is designed to detect a certain substance with the assumption that the substance indicates the presence of a harmful material, for example a flame photometer that detects phosphorous and reports the result on the assumption that the phosphorous came from a nerve agent. Yet others will report results based on a library match that can be 100% correct or 100% incorrect. All of these technologies are developed into instruments that require you to consider cross-sensitivity, false-positives, false-negatives, detection threshold, and assumptions on the use of the equipment. For these reasons, it is important to utilize more than one form of technology when characterizing hazards or identifying material; the use of three agreeing results assures high confidence. In order to use these tools to your benefit, you must understand the requirements and limitations of not only the technology, but the individual model of detector you are using.

This chapter is divided into three sections, one each for chemical, biological, and radiation detection technologies. As advances are made, new methods of combining existing and new technologies into a detection instrument can occur in ways not yet commercially available. While modern technology is able to provide new ways to analyze unknown material, the most basic tests are still important in a qualitative test scheme. Whenever an unknown material is discovered, a person skilled in qualitative analysis is the most important link between all forms of detection technology.

Chemical Detection Technology

Colorimetric Indicators

Colorimetric tests are based on the ability of one or more reagents to produce a visible change in color in the presence of targeted material. Colorimetric tests were a mainstay in chemical analysis from the 1700s and were used extensively through the 1940s. The 1950s saw the increased use of electronic instrumentation and shortly afterward, the advancement of several other technologies, including the advent of computerized analysis. Colorimetric tests still have a place in the identification of hazards posed by unknown material. The speed and simplicity of these tests make them dependable for field use as long as the operator assures the results are obtained within the parameters of the test.

Colorimetric tests, like all tests, have parameters and there may be more than one colorimetric reaction for detection of an analyte. When using a particular product, be sure you understand the limits of the test and the operating principle.

pH

Probably the most well known colorimetric test is for the determination of pH. Several types are available, but not are all applicable to field use. The concept of acidity was proposed in 1887 and the need for acid and base indicators became apparent.

Litmus

Litmus, a dye extracted from lichen, is a pH indicator. When the water-soluble litmus dye is dried on paper, it can be used to test pH. Red litmus paper turns blue to indicate a base and blue litmus paper turns red to indicate an acid. However, litmus indicates pH outside a range of approximately 4.3–8.1. Litmus indicator only classifies the material as an acid for materials with pH of 0–4.3 and as a base with pH of 8.1–14. Litmus will not produce a pH value.

Universal pH Indicator

pH test strips are available to determine pH throughout the range of 0–14. These strips may use a single paper square for the full pH range of 0–14 or a test strip may contain several paper squares of indicators specific to a subrange of pH. "Narrow range" pH test strips, which are limited to a lesser range of pH, should not be used for general pH determination. Figure 3.2 shows a universal pH indicator on a test strip indicating a strongly basic pH after application of ammonium hydroxide. If the small test strip is difficult to use due to extensive personal protective equipment such as in a hazardous materials emergency, the same indicator is available in a $1'' \times 12''$ strip as shown in Figure 3.3. This strip has an adhesive on the back so the strip can be attached to a pole or other object and testing can occur from a distance. These pH indicators are accurate over most of the scale. They are somewhat less accurate near the neutral area, but variation in this area is not a significant finding in field analysis. Strongly acidic or basic material will be clearly indicated. pH indicators will only work properly if water is present. Anhydrous material will not produce an accurate colorimetric result, so you must assure a small amount of water is present.

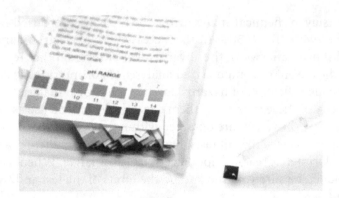

Figure 3.2 A universal pH indicator on a test strip indicating a strongly basic pH after application of ammonium hydroxide.

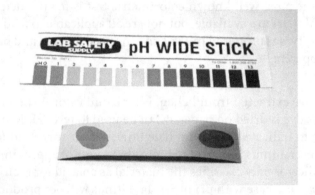

Figure 3.3 A universal pH indicator on a test strip indicating a strongly acidic pH on the left and a strongly basic pH on the right. pH indicators may be used to detect corrosive or caustic vapors as indicated by the "fog" of activated indicator surrounding the liquid ammonium hydroxide on the right.

pH indicators may also be used to detect strongly acidic or caustic vapors and gases if water is present. Some vapors may contain enough water to produce a result; partial detection may occur if humidity has dampened the pH test strip. When detecting vapors, use a pH strip moistened with distilled or deionized water and then allow suspect vapors to dissolve into the moist strip to produce a color. The premoistened pH test strip method is not as accurate since the water will partially fix the dye color before contact with the corrosive vapor or gas; however, it is dependable for the detection of strongly acidic and basic gases such as hydrogen chloride and ammonia.

Oxidizers
Colorimetric tests for oxidizing material are useful for field screening unknown material. These tests work by changing color based on oxidation potential or the presence of certain oxidizing compounds.

Potassium Iodide Starch Test Paper

Potassium iodide starch test paper indicates the presence of some oxidizers by changing from white through a range of blue to black depending on the concentration and oxidizing potential of oxidizing material in the sample. Potassium iodide (KI) and starch are present in the paper. Iodine will turn starch blue to black (Figure 3.4). The presence of a strong oxidizer will force iodide away from potassium and two iodide anions will bond to form iodine (I_2). Iodine is now available to bond with starch and a purple/black color is produced.

High concentration of a strong oxidizer will initially produce a black color, however, a secondary reaction can occur and the oxidizer will displace iodine from the starch and the test strip will turn white again (Figure 3.5).

Figure 3.4 Black color produced on a potassium iodide starch test strip.

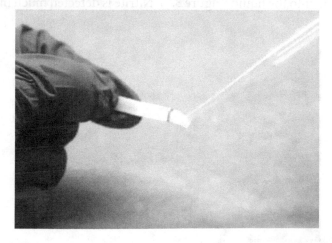

Figure 3.5 Black color bleached immediately to white on a potassium iodide starch test strip. Note the thin line of black formed at the leading edge of solution moving along the test strip.

Any production of blue/black color, no matter how brief, is a positive result for an oxidizer.

Some people have referred to potassium iodide starch test paper as "oxidizer" paper. It is not. The oxidizer must be able to displace iodide from potassium. Many oxidizers such as nitrate, perchlorate, and many organic peroxides are not able to displace iodide from potassium and are not detected. The addition of acid to the test paper does not remedy all these exceptions. While a positive result nearly always indicates an oxidizer, there are several oxidizers and conditions that do not force a color change on the white strip. Use with discretion.

Peroxide Test Strip

Many organic peroxides are not detectable by the potassium iodide starch method, although hydrogen peroxide will produce a black color. Detection of most peroxides requires a peroxide specific test paper (Figure 3.6). Peroxide test paper contains the enzyme peroxidase and a redox indicator. In the presence of peroxides and hydroperoxides, this enzyme will transfer an oxygen atom to the redox indicator and produce a blue oxidation product. Some test papers produce a brown color from a different indicator.

The strips detect both hydroperoxides and dialkyl peroxides. Some peroxidizable solvents, such as diethyl ether, may form polymeric peroxides that are either undetectable or detected only with reduced accuracy.

Nitrate/Nitrite Test Strip

Nitrate is a strong oxidizer and nitrite a weaker oxidizer that are not detected by the potassium iodide starch method. A test strip is commonly available that will detect both of these anions in water-based solutions by simply dipping the strip into the liquid (Figure 3.7). Nitrite is detected much more easily

Figure 3.6 Positive result on peroxide test paper from exposure to 3% hydrogen peroxide.

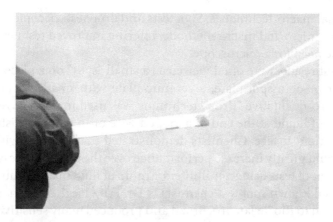

Figure 3.7 Detection of nitrate in an aqueous solution by a nitrate/nitrite test strip.

than nitrate and interferes with nitrate detection; nitrite is not expected to be as oxidizing as nitrate when encountered as an unknown substance. If nitrite is present, some test strips include a reagent (amidosulfuric acid) that will destroy nitrite and unmask nitrate. Since nitrite is not encountered as often as nitrate, an assumption can be made in the field that a positive result on the nitrate/nitrite test paper is strongly suggestive of nitrate. A practical field detection limit has been suggested as 10 mg/l nitrate and 5 mg/l nitrite.

Perchlorate Test Solution
Perchlorate is an oxidizer used in rocket fuel and explosives and is not detected by the general potassium iodide starch method. Concentrated perchlorate solutions are dangerously unstable. A simple colorimetric test for perchlorate involves the use of a dilute methylene blue solution, which will turn purple in the presence of perchlorate. The test may be performed in a test tube or other small open container, or by spraying the solution on the suspect material. Violet color will develop from the deep blue within a few seconds. The test must be water based and the test is not stable if dried on paper.

A practical detection limit has been reported as 0.001 M perchlorate using a 0.3% aqueous solution of methylene blue. Upon contact with methylene blue, chromates will produce a brown-red color; some cyanide compounds will produce a green color.

Spot and Drop Tests for Toxic Material
Spot tests were developed in the early part of the twentieth century when chemists worked with very small, but visible, amounts of reagents. Chemists found tests could be more specific and sensitive if conducted on a smaller scale. Microchemistry utilized a microscope to view results instead of

conventional macro-techniques. Spot tests and drop tests occupied a position between the micro and macro methods, offering improved test performance without the need for a microscope.

Spot test results were often viewed as a small "spot" on a piece of treated filter paper or on a spot plate, a ceramic plate with small indentations. A similar micro-qualitative "drop" technique was used to observe results in a single drop of liquid at the end of a looped wire or in a micro-dish, a small, curved reaction surface. Chemists demonstrated that working with smaller volumes could greatly increase performance over the crude and less sensitive macro format of tests using a milliliter or more of fluid in a test tube. Among the most well known spot test chemists is Dr. Fritz Fiegl, a German chemist who relocated to Rio de Janeiro, Brazil and produced many sensitive and useful tests during the 1930s and the 1960s.

Several metals can cause long term toxicity hazards. Contact with these metals should be avoided due to the inability of the body to eliminate them. The EPA requires waste coding to reflect certain hazards. Toxic metals specifically mentioned include arsenic, barium, cadmium, chromium, lead, mercury, selenium, and silver. The EPA also describes acceptable test methods, most of which are laboratory based and require specialized instrumentation.

Several simple and quick field tests are given below for screening unknown material for these especially toxic metals. Many other tests are found in the literature, but have not been included here because they would be impractical for field use; however, they might be applicable to some laboratory analysis schemes. If developing your own tests from the literature, be sure to research the reagents for health hazards. Many old tests use reagents now considered too dangerous to be used for this purpose. One such reagent, benzidine, was widely useful as a colorimetric reagent, but has been found to be a potent carcinogen. These tests are included for their simplicity and portability; the tests are useful field screens but are not sensitive enough in most cases to supplant EPA-approved methods.

Cyanide
Cyanide can be present in various forms in water. At high doses, loosely bound forms of cyanide such as potassium or sodium cyanide inhibit cellular respiration and, in some cases, can result in death. In water under ambient conditions, cyanide evolves from aqueous hydrogen cyanide, sodium cyanide, potassium cyanide, and other metal or ionic salts where cyanide is released when dissolved in water. Heavier cyanide complexes formed by some metals are bound tightly, and the addition of acid is necessary to liberate toxic-free cyanide ion, which can leave concentrated solutions as hydrogen cyanide gas. Because disassociation of the free cyanide ion is unlikely under ambient conditions, the heavier salts are considered much less toxic than simple cyanide salts. Field test strips for cyanide are used to detect free cyanide and can be used to easily detect the most toxic cyanide salts.

There are a few tests for cyanide, but the most convenient test converts cyanide in water to cyanogen chloride in the presence of isonicotinic and barbituric acids and produces a blue color change that can be detected visually. A result of 0.1 mg/l to >10 mg/l can be obtained visually in approximately 1 minute. An optical reader can provide lower detection if necessary, but visual results are adequate to quickly characterize the hazards of a sample.

Remember that using acid to ionize tightly bound metal cyanides will evolve toxic hydrogen cyanide gas as the pH shifts to acid. Assure adequate ventilation and respiratory protection.

Sulfide
The sulfides of most metals are insoluble in water. Those that are water soluble are the alkaline earth and alkali metal sulfides, aluminum sulfide, and chromium sulfide. Sulfides exist in aqueous solution with equilibrium determined by pH. Near neutral pH hydrogen sulfide ion (HS$^-$) is predominant and in basic conditions sulfide ion (S^{2-}) is predominant. In acidic conditions, sulfides exist as dissolved toxic hydrogen sulfide gas (H$_2$S), which will bubble out of the solution as concentration increases. Increasing acidity will produce greater amounts of hydrogen sulfide gas.

Several reagents can detect sulfides in low concentrations. The most practical is lead acetate dried onto filter paper. Lead acetate will produce a brown to black color in the presence of sulfide in liquid. It is important to note that lead acetate test paper is not adequately sensitive to quickly detect hydrogen sulfide in air. Even at very low concentrations of hydrogen sulfide in basic solutions, lead acetate test strips can be used to detect the presence of 5–10 ppm sulfide. You can also test a weak solution by adding acid and testing the air above the sample with a colorimetric air monitoring tube or an air monitor with an electrochemical sensor while controlling any generation of hydrogen sulfide gas with adequate ventilation while using respiratory protection.

Chlorine
Chlorine in water solutions exists in equilibrium dependent on pH. In basic solutions, hypochlorite ion is predominant, as in household bleach. In neutral to acidic solutions, hypochlorous acid and free chlorine are predominant. Chlorine gas will bubble from a water solution that is mixed with enough acid.

Most chlorine test strips are designed to detect hypochlorite ion for determination of bleach in disinfecting solutions, pool water, and drinking water. The tests are often pH dependent and some require a buffer. The test strips are not necessarily a measure of chlorine, but will react to strong oxidizers. In some tests, chloride interferes. If a chlorine gas-producing solution is suspected, you could use a more specific test such as adding acid to the solution and drawing a gas sample through a chlorine specific colorimetric sampling tube.

Fluoride

Soluble and insoluble fluoride material may be detected by use of zirconium alizarin.

The zirconium alizarin complex in water is violet and is in equilibrium. The dissociated component is yellow, but the equilibrium favors the violet color. When zirconium ions are removed from the solution by the addition of fluoride ions, the yellow color is favored. The reagent can be prepared by adding zirconium chloride and alizarin in approximately equal weights in water to form a 1% solution for liquid testing. The reagent is stable and may be dried on paper from a 5% solution. Fluoride test strips are available commercially.

Phosphates, arsenates, oxalates, and large quantities of sulfate will interfere with this test. If one or more of these materials are present, the sample must be prepared by precipitating the interfering material (or fluoride with calcium) and removing it from the test solution.

Water

Water is certainly not a highly toxic material, but knowing if water is present can help determine characteristics of a liquid. A simple method of determining the presence of water at more than 10% concentration in a liquid is the addition of a piece of Alka-Seltzer® or a generic equivalent (Figure 3.8). This medicine is a dry combination of citric acid and sodium carbonate. The presence of water will allow the acid and base to react and evolve carbon dioxide gas bubbles from the solution. This test can be used to determine if a polar organic liquid or other fluid contains more than about 10% water.

Figure 3.8 Alka-Seltzer effervescing in an aqueous solution of 70% isopropyl alcohol.

Spot and Drop Tests for Metals

Arsenic

Arsenic cations exist as trivalent (As^{+3}) and pentavalent (As^{+5}) forms. Arsenic anions also exist as trivalent arsenite (AsO_3^{-3}) and pentavalent arsenate (AsO_4^{-3}) forms.

Arsenate (AsO_4^{-3}) can be detected by first adjusting the pH to neutral and then adding silver nitrate (1.7% w/v). A chocolate-brown precipitate indicates silver arsenate. Arsenite (AsO_3^{-3}) in the same process will yield a yellow precipitate of silver arsenite, but it is not very sensitive.

Trivalent arsenic (As^{+3}) can be detected as described by Bettendorf. Place two drops of concentrated hydrochloric acid on a spot plate and add a drop of the unknown solution. Add a drop of stannous chloride (saturated solution in concentrated hydrochloric acid). A brown color forms in less than a minute.

There is no specific spot test for pentavalent arsenic, but if arsenic in any form can be reduced to arsine, a sensitive test described by Gutzeit may be used. Place the unknown in a test tube and add a few granules of zinc followed by a few drops of dilute sulfuric acid. Arsenic reduced to arsine gas will leave the solution and rise to the top of the test tube. A strip of filter paper moistened with silver nitrate solution and placed over the opening of the tube will produce a gray or yellow spot due to formation of a double salt, $AsAg_3 \cdot AgNO_3$.

Barium

Barium was traditionally detected by forming a yellow precipitate with the addition of potassium chromate in water (10% w/v). Lead also produces a yellow precipitate with chromate.

Barium may be detected by a fresh solution of a sodium rhodizonate (1% w/v) in water. Barium will produce a red color, as will strontium but not calcium. In a torch flame, barium will produce a yellow-green flame and strontium will produce a road flare red flame.

Cadmium

Accurate cadmium detection is difficult in a field setting. Feigl described a very sensitive method, but several steps are involved using a key reagent, a,a'-dipyridil (CAS 366-18-7), also known as 2,2'-bipyridine. Ammoniacal solution of cadmium with the addition of ferrous a,a'-dipyridil and iodide ions will immediately produce a red precipitate.

The procedure involves first making ferrous a,a'-dipyridil by mixing a ferrous salt such as ferrous sulfate (0.146 g) and a,a'-dipyridil hydrochloride (0.25 g) in 50 ml water followed by the addition of an iodide source such as potassium iodide (10 g). Filter the solution if it is cloudy. Next, dissolve solid cadmium with nitric acid, dry the result, and take it up with ammonium hydroxide solution. Filter the solution only if it is very cloudy with a precipitate; light precipitate is acceptable. Simultaneously add a drop of each solution to a filter paper and watch for a red fleck or ring, indicating cadmium.

A less accurate and less sensitive method is the addition of the a,a'-dipyridil solution directly to the unknown substance. If cadmium ion is present, a red solution will occur. If cadmium is bound, it must be ionized for the test to detect it.

Chromium

Chromium exists in several forms. Only the trivalent (Cr^{+3}) and hexavalent cations (Cr^{+6}) and anions chromate (CrO_4^{-2}) and dichromate ($Cr_2O_7^{-2}$) are common. Chromate solution and lead nitrate solution will form a yellow precipitate of lead chromate.

Chromate and dichromate may be detected by diphenyl carbazide depending on the oxidizing action of the anion. That is, chromate must be oxidized to dichromate to be detected. This oxidation may be accomplished by adding sodium hydroxide and sodium peroxide to the unknown solution. Mix a saturated diphenyl carbazide solution in 90% alcohol and then saturate with potassium thiocyanate. Add a few grains of potassium iodide and mix well. Add a drop of this solution to a reaction paper. Next add a drop of the sample solution and then add a drop of dilute sulfuric acid. A violet color appears if dichromate from the original chromium source is present. (Red indicates magnesium.)

Chromium III (Cr^{+3}) forms a pale green solution in ammonium hydroxide. If this solution is oxidized with sodium peroxide or another suitable oxidizer, the solution turns yellow when Cr^{+3} converts to chromate (CrO_4^{-2}). The diphenyl carbizide test above will confirm chromium.

Chromium VI solutions are orange and oxidizing. An orange solid or liquid with a positive result on a potassium iodide starch test strip should be presumed to contain chromium VI.

Lead

Lead salt solution can be detected by the addition of potassium chromate in water (10% w/v), which will cause the precipitation of yellow lead chromate. Barium also produces a yellow precipitate.

Mercury

Mercury ions may exist in two cationic forms, mercurous (Hg_2^{++}) and mercuric (Hg^+). A stannous chloride solution added to mercurous or mercuric salt solution will immediately form a white precipitate that quickly turns gray and then black. The black is due to the formation of metallic mercury. These compounds are not precipitated by chloride.

Alternatively, concentrated potassium nitrate solution added to mercurous salt solution will form a black spot on paper. In the presence of colored ions, a brown spot will form that can be washed off to reveal the black mercury spot.

Mercuric salt solution can be detected by adding a grain of ammonium thiocyanate and then a grain of cobalt acetate. A blue color will form due to the formation of a cobalt mercury double salt.

Selenium

Free selenium exists in several allotropic modifications and is difficult to detect in all forms. The presence of free sulfur causes interference in colorimetric tests. There is a reference to the addition of thallous sulfide causing the formation of a black color, but a practical and reliable test for selenium is not readily apparent.

Silver

Silver salt solutions react with soluble chromates to produce a red-silver chromate precipitate.

To detect silver, place a drop of the unknown liquid on a spot plate. Add a drop of ammonium carbonate solution (10 g into 10 ml of concentrated ammonium hydroxide and 50 ml of water and then dilute with more water to 100 ml). Transfer the clear liquid from the precipitate that forms in the spot plate to another spot and add a drop of potassium chromate solution (10% w/v). Red silver chromate appears as a positive result.

Flame Tests

Unknown material subjected to a flame may produce recognizable characteristics, most notably flame color. Visual detection of flame color is dependent on concentration of the material, presence of material that can mask the flame color, and the temperature of the flame.

Since flame color can be masked by even small amounts of some common contaminants, these flame tests can be used to confirm the presence of a material, but lack of a positive result cannot be used to rule out a material.

Detection of flame color is subjective and is based on the observer's skill and practice. The variability of flame temperature can be diminished if the same burner and fuel are used each time. Natural gas is available in a fixed facility, but propane is more likely to be used in portable applications. Butane is not reliable in cold weather applications. Flammable liquids may be used on a heat resistant dish, but many substances will not be taken up in the flame. Additionally, most of the flammable liquid must burn away before solid material is heated to the point where it will produce a colored flame. Acetlylene-oxygen is too intense for use in flame tests. A portable propane torch supplied by a small cylinder of LPG is the most desirable choice overall.

The metal used to hold the sample in the flame must be cleaned by heating it in the flame until no significant flame discoloration occurs. If the metal is not "cleaned" within 20–30 seconds, the metal itself is not acceptable for the test. Some metals will produce a large orange flame color that masks other results. Most steel will not interfere, so some wires, hairpins, paperclips, and spatulas will be acceptable, but each must be subjected to the torch flame before each use to be sure.

Sodium is an abundant contaminant that will produce a persistent and large orange color that will mask all other colors, even if the sodium is in low concentration.

The test is performed by placing a small amount of the material on a piece of metal and holding it in the hottest part of the propane torch flame, which is the tip of the light blue area near the torch head. Look for a change in color in the propane flame downstream from the sample (Figure 3.9; Table 3.1). Remember the heat that is traveling toward your hand through the tool that is holding the sample in the flame.

Figure 3.9 Potassium flame color in propane torch flame. Note the position of the sample in the hottest area, which is the tip of the inner, lighter blue part of the flame.

Table 3.1   Elemental Color in Propane Flame

Element	Flame Color
Arsenic	Blue
Barium	Yellow-green
Boron	Green
Calcium	Orange-red
Copper	Green
Lead	Gray-blue
Lithium	Red
Potassium	Light purple
Sodium	Yellow-orange

Copper Wire (Beilstein)

Feigl reports detection of organic compounds containing halogens (Cl, I, Br), cyanide, and thiocyanide with a copper wire heated in a torch flame (the test does not work on organofluorine material). A thin copper wire or a thicker wire hammered flat is cleaned in the torch flame while a coating of copper oxide forms (Figure 3.10). The sample is applied to the cooled copper and placed in the flame. The test may also be performed on vapors that are introduced through the air intake of the torch while the hot copper wire is held in the flame. The previously mentioned material will combine with the copper oxide to produce volatile copper halides that will produce a characteristic green or blue flame (copper fluoride is not volatile). Interference is caused by some nitrogen-containing material such as urea, pyridine, and oxyquinolines, which will form volatile copper cyanide and produce a green flame color. The presence of some acids, especially halogen acids, produces a green flame color.

Test Strips

Many colorimetric test strips have been developed for specific applications, such as the detection of heavy metals, anions, biological material, etc. Most of these test strips respond selectively. Some test strips have masking agents in the reaction zone to suppress various interferences. Interferences cannot be avoided in every case, especially if the concentration is high. Check with the product manufacturer for the parameters of the test and a list of known interfering substances. Due to these variables, test strips should not be used

Figure 3.10 Chlorinated solvents are detected with a copper wire in the flame when vapors are drawn into the torch air intake. Low vapor pressure liquids may need to be heated slightly to increase vapor production.

quantitatively except in the broadest sense when testing unknown material. Table 3.2 is a partial list of test strips available from global suppliers.

Most test strips will carry an expiration date. They should be stored in a dry, cool place. Some will need to be refrigerated. Always refer to the manufacturer's instructions on specific use of the test.

Explosive Material Tests

Colorimetric tests for the detection of explosive material have been in existence for over a century. Many of the early reagents were not stable in storage, were explosive, or later found to be carcinogenic. Newer discoveries have provided field colorimetric tests that are suitable for use for a wide variety of explosive materials.

The "hammer and anvil" test has long been used as a field method to characterize a material as shock sensitive. To determine if a substance is shock sensitive, place a milligram of sample (equivalent to 5–10 grains of table salt) on a hard, inert surface and give it a good whack with a hammer. Since you are suspecting a crisp and powerful explosion, use high-quality material for the hammer and anvil that will not fragment when an explosive wave passes through it. Perform this test in a manner that controls any material that may be ejected when struck. Preferably, you should use a device that you can operate remotely so you are removed from the area of any possible fragmentation. Wear appropriate protective equipment such as a face shield and ear protection.

Drop-Ex Plus™ (The Mistral Group) is a set of sequential colorimetric tests for the detection of chemical structures present in several explosive

Table 3.2 Commercially Available Colorimetric Tests

Alkaline phosphatase	Cobalt	Permanganate
Aluminum	Copper	Peroxide
Ammonia	Cyanide	Phosphate
Ammonium	Formaldehyde	Potassium
Antimony	Hydrogen peroxide	Potassium
Arsenic	Iodine	Protein residues
Arsine	Iron	Quaternary ammonium
Ascorbic acid	Lead	Silver
Bismuth	Manganese	Sulfate
Borates	Molybdenum	Sulfite
Bromine	Nickel	Tin
Calcium	Nitrate	Water in organic solvent
Carbonate hardness	Nitrite	Zinc
Chloride	Nitrous acid	Zirconium
Chlorine	Ozone	
Chromate	Peracetic acid	

materials. The tests use an absorbent paper wipe or some other substrate to collect a very small sample of suspected explosive material. Spray bottles contain mixtures of pretreatment chemicals, colorimetric reagents, solvents, buffers, etc. The sprays are applied to the wipe sample in a specified order since sequential tests utilize reactions from previous sprays. The level of sensitivity is 20 nanograms or less. The tests detect certain organic structures, functional groups, and degradation products; positive results are possible from interfering material.

Drop-Ex Plus detects:

- Nitroaromatics such as TNT, tetryl, TNB, DNT, picric acid, and its salts
- Nitrate esters and nitramines such as dynamite, nitroglycerine, RDX, C4, PETN, Semtex H, nitrocellulose, smokeless powder, and many of the plastic explosives
- Inorganic nitrates such as ANFO, black powder, gunpowder, potassium nitrate, and ammonium nitrate
- Chlorates and bromates

Peroxides Such as TATP and HMTD

Another example of a field screening test kit for explosive material is ELITE™ Model EL100, EL200 and EL 300 (Field Forensics, Inc.). This kit uses a wipe sample that is slid into a card. Reagents in glass ampoules are crushed in a specified order and colorimetric results are produced by certain organic structures, functional groups, and degradation products; positive results are possible from interfering material. Some tests are heated with a lighter or battery-operated heater. These tests use chemistry different than Drop-Ex Plus, so the list of detectable material is also different.

The three basic ELITE tests, EL100, EL200, and EL300 can screen for many explosive materials:

- EL100 tests for nitro aromatic, nitro aliphatic, and nitrate-based explosives and precursors, such as TNT, DNT, TATB, RDX, HMX, EGDN, PETN, NG, NC, ANFO, ANS, and black powder.
- EL200 tests for chlorate, peroxide, and perchlorate-based HME and precursors, such as TATP, HMTD, MEKP, flash powders, Armstrong's mixture, Poor Man's C4, potassium chlorate mixtures.
- EL300 tests for nitro aromatic, nitro aliphatic, peroxide, chlorate and nitrate-based explosives and precursor such as TNT, DNT, TATB, RDX, HMX, EGDN, PETN, NG, NC, ANFO, ANS, TATP, HMTD, MEKP, black powder, flash powders, Armstrong's mixture, Poor Man's C4.

Follow-on tests can be used to specifically identify many explosives and precursors. Both test kits detect trace amounts of a wide range of explosive material. A trained operator should consult the kit instructions before use.

Drug Screen

Unlabeled material found under suspicious circumstances might be assumed to be an illicit drug. It is not unreasonable to assume a law enforcement officer might perform a drug field test. The U.S. Department of Justice has set a National Institute of Justice Standard 0604.01, *Color Test Reagents/Kits for Preliminary Identification of Drugs of Abuse*. This standard clearly outlines approved reagents and test methods for the identification of some drugs. Understandably, common materials that might be mistaken for a drug may also present a set of findings in the test matrix (Table 3.3). The test matrix is reproduced here so that material that has been drug screened might provide additional clues as to hazards present. Table 3.4 describes the colors produced in the tests and Table 3.5 provides instructions for mixing the test reagents.

The test reagents can be used individually to rule in or rule out a substance based on the data presented by the drug screen. Detection limits are typically no more than a few hundred micrograms. For more information refer to Color Test Reagents/Kits for Preliminary Identification of Drugs of Abuse, NIJ Standard-0604.01. Individual reagents from this drug screen are often packaged into presumptive field test kits as shown in Figure 3.11.

According to the results in Table 3.3, aspirin will not produce a colored result with the addition of reagent A.8, ferric chloride. Aspirin is identified by the drug screen through negative results with all reagents except positive results with reagents A.4, A.5, and A.9. Aspirin can also be identified with a clear aqueous solution of ferric nitrate that will turn deep purple in contact with aspirin.

Alkaloids are derivatives of amino acids. Most alkaloids are detected by cobalt thiocyanate, Reagent A.1 in Table 3.5. Alkaloids include the drugs ending in "-caine" such as cocaine, lidocaine, and xylocaine and also include strychnine, morphine, atropine, ephedrine, caffeine, ricinine, and many others. Cobalt thiocyanate will react with other material, so a positive result must be confirmed with other test methods or use of a matrix similar to the drug screen.

M256A1 Kit

The M256A1 Kit is used by the U.S. military and allies to detect chemical warfare agent liquids and vapors. The kit contains instructions, M8 paper for use with liquid samples, and a sampler detector for use with vapors (Figure 3.12).

M8 is used by the U.S. military to detect G- and V-nerve and H-blister agents in the field under combat conditions. When an "unnatural" liquid is found in a "natural" setting of battlefield conditions it is presumed to be

Table 3.3 Screen for Drugs and Commonly Mistaken Material

(+) Indicates that a color reaction occurs[a]	Reagent											
	A.1	A.2	A.3	A.4	A.5	A.6	A.7	A.8	A.9	A.10	A.11	A.12
Acetominophen	−	−	−	+	−	+	−	+	−	−	−	−
Alprazolam												
Aspirin	−	−	−	+	+	−	−	−	+	−	−	−
Baking soda	−	−	−	−	−	−	−	+	−	−	+	−
Brompheniramine maleate	+	−	−	+	−	−	−	−	−	−	−	−
Chlordiazepoxide HCl	+											
Chlorpromazine HCl	+	−	−	+	+	+	−	+	+	+	−	−
Contac®	−	−	−	+	−	−	−	−	+	+	−	−
Diazepam												
Doxepin HCl	+	−	−	+	+	+	−	−	+	+	−	−
Dristan®	−	−	−	+	+	+	−	+	+	+	−	−
Ephedrine HCl	+											
Exedrine®	−	−	−	+	+	+	−	+	+	+	+	−
Hydrocodone tartrate	+	−	−	−	−	−	−	−	+	−	−	
Mace²	−	−	+	+	+	+	−	−	+	+	−	
Meperidine HCl	+	−	−	−	|	−	−	−	−	−	−	
Methaqualone	−	−	−	+	−	−	−	−	−	−		
Methylphenidate HCl	+	−	−	|	+							+
Nutmeg[b]	−	−	+	−	−	−	−	−	−	−	+	−
Phencyclidine HCl	+											
Propoxyphene HCl	+	−	−	+	+	−	−	−	+	+	−	−
Pseudoephedrine HCl	+											
Quinine HCl	+	−	−	+	−	−	−	−	−	−	−	−
Salt	−	−	−	+	−	−	−	−	−	−	−	−
Sugar	−	−	−	−	+	−	−	−	−	−	−	−
Tea[b]	−	−	+	−	−	−	−	−	−	−	+	−
Tobacco											+	−

[a] Substances that gave no colors with these reagents are: D-galactose, glucose, mannitol, oregano, rosemary, and thyme.

[b] Tea, mace, and nutmeg may interfere with the Duquenois test but not the Duquenois-Levine modified test (A.3).

Source: Table reproduced from Color Test Reagents/Kits for Preliminary Identification of Drugs of Abuse, NIJ Standard-0604.01, U.S. Department of Justice.

Table 3.4   Final Colors Produced by Reagents A.1–A.12 with Various Drugs and Other Substances

Reagent	Analyte	Solvent	Color
A.1	Benzphetamine HCl	Chloroform	Brilliant greenish blue
A.1	Brompheniramine maleate	Chloroform	Brilliant greenish blue
A.1	Chlordiazepoxide HCl	Chloroform	Brilliant greenish blue
A.1	Chlorpromazine HCl	Chloroform	Brilliant greenish blue
A.1	Cocaine HCl	Chloroform	Strong greenish blue
A.1	Diacetylmorphine HCl	Chloroform	Strong greenish blue
A.1	Doxepin HCl	Chloroform	Brilliant greenish blue
A.1	Ephedrine HCl	Chloroform	Strong greenish blue
A.1	Hydrocodone tartrate	Chloroform	Brilliant greenish blue
A.1	Meperidine HCl	Chloroform	Strong greenish blue
A.1	Methadone HCl*	Chloroform	Brilliant greenish blue
A.1	Methylphenidate HCl	Chloroform	Brilliant greenish blue
A.1	Phencyclidine HCl	Chloroform	Strong greenish blue
A.1	Procaine HCl*	Chloroform	Strong greenish blue
A.1	Propoxyphene HCl*	Chloroform	Strong greenish blue
A.1	Pseudoephedrine HCl	Chloroform	Strong greenish blue
A.1	Quinine HCl	Chloroform	Strong blue
A.2	Amobarbital	Chloroform	Light purple
A.2	Pentobarbital*	Chloroform	Light purple
A.2	Secobarbital*	Chloroform	Light purple
A.3	Mace[e] crystals		Strong reddish purple to very light purple
A.3	Nutmeg extract		Pale reddish purple to light gray purplish red
A.3	Tea extract		Light yellow-green
A.3	THC*	Ethanol	Gray purplish blue to light purplish blue to deep purple
A.4	Acetaminophen	Chloroform	Moderate olive
A.4	Aspirin powder		Grayish olive green
A.4	Benzphetamine HCl*	Chloroform	Brilliant yellow-green
A.4	Brompheniramine maleate	Chloroform	Strong orange
A.4	Chlorpromazine HCl	Chloroform	Dark olive
A.4	Cocaine HCl*	Chloroform	Deep orange-yellow
A.4	Codeine*	Chloroform	Dark olive
A.4	Contac powder		Strong yellow
A.4	d-Amphetamine HCl*	Chloroform	Moderate bluish green
A.4	d-Methamphetamine HCl*	Chloroform	Dark yellowish green
A.4	Diacetylmorphine HCl*	Chloroform	Moderate reddish brown
A.4	Dimethoxy-meth HCl	Chloroform	Dark olive brown
A.4	Doxepin HCl	Chloroform	Dark reddish brown

(Continued)

Table 3.4 (Continued)   Final Colors Produced by Reagents A.1–A.12 with Various Drugs and Other Substances

Reagent	Analyte	Solvent	Color
A.4	Dristan powder		Grayish olive
A.4	Exedrine powder		Dark olive
A.4	Mace^e crystals		Moderate olive green
A.4	MDA HCl	Chloroform	Bluish black
A.4	Mescaline HCl*	Chloroform	Dark yellowish brown
A.4	Methadone HCl	Chloroform	Dark grayish blue
A.4	Methaqualone	Chloroform	Very orange-yellow
A.4	Methylphenidate HCl	Chloroform	Brilliant orange-yellow
A.4	Morphine monohydrate*	Chloroform	Dark grayish reddish brown
A.4	Opium*	Chloroform	Dark brown
A.4	Oxycodone HCl	Chloroform	Dark greenish yellow
A.4	Procaine HCl	Chloroform	Deep orange
A.4	Propoxyphene HCl	Chloroform	Dark reddish brown
A.4	Quinine HCl	Chloroform	Deep greenish yellow
A.4	Salt crystals		Strong orange
A.5	Aspirin powder		Deep red
A.5	Benzphetamine HCl*	Chloroform	Deep reddish brown
A.5	Chlorpromazine HCl	Chloroform	Deep purplish red
A.5	Codeine*	Chloroform	Very dark purple
A.5	d-Amphetamine HCl*	Chloroform	Strong reddish orange to dark reddish brown
A.5	d-Methamphetamine HCl*	Chloroform	Deep reddish orange to dark reddish brown
A.5	Diacetylmorphine HCl*	Chloroform	Deep purplish red
A.5	Dimethoxy-meth HCl	Chloroform	Moderate olive
A.5	Doxepin HCl	Chloroform	Blackish red
A.5	Dristan powder		Dark grayish red
A.5	Exedrine powder		Dark red
A.5	LSD chloroform		Olive black
A.5	Mace^e crystals		Moderate yellow
A.5	MDA HCl*	Chloroform	Black
A.5	Meperidine HCl	Chloroform	Deep brown
A.5	Mescaline HCl*	Chloroform	Strong orange
A.5	Methadone HCl	Chloroform	Light yellowish pink
A.5	Methylphenidate HCl	Chloroform	Moderate orange-yellow
A.5	Morphine monohydrate*	Chloroform	Very deep reddish purple
A.5	Opium* powder		Dark grayish reddish
A.5	Oxycodone HCl*	Chloroform	Pale violet
A.5	Propoxyphene HCl	Chloroform	Blackish purple
A.5	Sugar crystals		Dark brown

(Continued)

Table 3.4 (Continued)   Final Colors Produced by Reagents A.1–A.12 with Various Drugs and Other Substances

Reagent	Analyte	Solvent	Color
A.6	Acetaminophen	Chloroform	Brilliant orange-yellow
A.6	Codeine*	Chloroform	Light greenish yellow
A.6	Diacetylmorphine HCl*	Chloroform	Pale yellow
A.6	Dimethoxy-meth HCl	Chloroform	Very yellow
A.6	Doxepin HCl	Chloroform	Brilliant yellow
A.6	Dristan powder		Deep orange
A.6	Exedrine powder		Brilliant orange-yellow
A.6	LSD	Chloroform	Strong brown
A.6	Mace^e crystals		Moderate greenish yellow
A.6	MDA HCl	Chloroform	Light greenish yellow
A.6	Mescaline HCl*	Chloroform	Dark red
A.6	Morphine monohydrate*	Chloroform	Brilliant orange yellow
A.6	Opium* powder		Dark orange yellow
A.6	Oxycodone HCl	Chloroform	Brilliant yellow
A.7	LSD*	Chloroform	Deep purple
A.8	Acetaminophen	Methanol	Dark greenish yellow
A.8	Baking soda powder		Deep orange
A.8	Chlorpromazine HCl	Methanol	Very orange
A.8	Dristan powder		Moderate purplish blue
A.8	Exedrine powder		Moderate purplish blue
A.8	Morphine monohydrate*	Methanol	Dark green
A.9	Aspirin powder		Grayish purple
A.9	Chlorpromazine HCl	Chloroform	Very deep red
A.9	Codeine*	Chloroform	Very dark green
A.9	Contac powder		Moderate olive brown
A.9	Diacetylmorphine HCl*	Chloroform	Deep purplish red
A.9	Dimethoxy-meth HCl	Chloroform	Very yellow-green
A.9	Doxepin HCl	Chloroform	Deep reddish brown
A.9	Dristan powder		Light bluish green
A.9	Exedrine powder		Brilliant blue
A.9	LSD chloroform		Moderate yellow-green
A.9	Mace^e crystals		Light olive yellow
A.9	MDA HCl*	Chloroform	Greenish black
A.9	Morphine monohydrate*	Chloroform	Deep purplish red
A.9	Opium* powder		Brownish black
A.9	Oxycodone HCl	Chloroform	Strong yellow
A.9	Propoxyphene HCl	Chloroform	Dark grayish red
A.9	Sugar crystals		Brilliant yellow
A.10	Chlorpromazine HCl	Chloroform	Blackish red

(Continued)

Table 3.4 (Continued) Final Colors Produced by Reagents A.1–A.12 with Various Drugs and Other Substances

Reagent	Analyte	Solvent	Color
A.10	Codeine*	Chloroform	Very dark bluish green
A.10	Contac powder		Moderate olive brown
A.10	Diacetylmorphine HCl*	Chloroform	Deep bluish green
A.10	Dimethoxy-meth HCl	Chloroform	Dark brown
A.10	Doxepin HCl	Chloroform	Very dark red
A.10	Dristan powder		Light olive brown
A.10	Exedrine powder		Dark grayish yellow
A.10	Hydrocodone tartrate	Chloroform	Dark bluish green
A.10	LSD chloroform		Greenish black
A.10	Mace[5] crystals		Dark grayish olive
A.10	MDA HCl*	Chloroform	Very dark bluish green
A.10	Mescaline HCl*	Chloroform	Moderate olive
A.10	Morphine monohydrate[d]	Chloroform	Very dark bluish green
A.10	Nutmeg leaves		Brownish black
A.10	Opium* powder		Olive black
A.10	Oxycodone HCl	Chloroform	Moderate olive
A.10	Propoxyphene HCl	Chloroform	Deep reddish brown
A.10	Sugar crystals		Brilliant greenish yellow
A.11	Baking soda powder		Light blue
A.11	Exedrine powder		Light green
A.11	Pentobarbital*	Chloroform	Light purple
A.11	Phenobarbital*	Chloroform	Light purple
A.11	Secobarbital*	Chloroform	Light purple
A.11	Tea leaves		Moderate yellow-green
A.11	Tobacco leaves		Moderate yellowish green
A.12	d-Methamphetamine HCl*	Chloroform	Dark blue
A.12	Dimethoxy-meth HCl*	Chloroform	Deep blue
A.12	MDMA HCl	Chloroform	Dark blue
A.12	Methylphenidate HCl	Chloroform	Pale violet

* Usual kit reagent for that particular drug.
[a] Aqueous phase.
[b] Aqueous phase after chloroform extraction.
[c] Chloroform phase.
[d] Not extracted into chloroform.
[e] 2-Chloroacetophenone.

Source: Adapted from Color Test Reagents/Kits for Preliminary Identification of Drugs of Abuse, NIJ Standard–0604.01.

Table 3.5   Drug Test Reagent Ingredients and Procedures

A.1 Cobalt Thiocyanate	Dissolve 2.0 g of cobalt (II) thiocyanate in 100 ml of distilled water.
A.2 Dille-Koppanyi reagent, modified	Solution A: Dissolve 0.1 g of cobalt (II) acetate dihydrate in 100 ml of methanol. Add 0.2 ml of glacial acetic acid and mix. Solution B: Add 5 ml of isopropylamine to 95 ml of methanol. Procedure: Add 2 volumes of solution A to the drug, followed by 1 volume of solution B.
A.3 Duquenois-Levine reagent, modified	Solution A: Add 2.5 ml of acetaldehyde and 2.0 g of vanillin to 100 ml of 95% ethanol. Solution B: Concentrated hydrochloric acid. Solution C: Chloroform. Procedure: Add 1 volume of solution A to the drug and shake for 1 min. Then add 1 volume of solution B. Agitate gently, and determine the color produced. Add 3 volumes of solution C and note whether the color is extracted from the mixture to A and B.
A.4 Mandelin reagent	Dissolve 1.0 g of ammonium vanadate in 100 ml of concentrated sulfuric acid.
A.5 Marquis reagent	Carefully add 100 ml of concentrated sulfuric acid to 5 ml of 40% formaldehyde (v:v, formaldehyde:water).
A.6 Nitric acid	Concentrated nitric acid.
A.7 Para-Dimethylaminobenzaldehyde (p-DMAB)	Add 2.0 g of p-DMAB to 50 ml of 95% ethanol and 50 ml of concentrated hydrochloric acid.
A.8 Ferric chloride	Dissolve 2.0 g of anhydrous ferric chloride or 3.3 g of ferric chloride hexahydrate in 100 ml of distilled water.
A.9 Froede reagent	Dissolve 0.5 g of molybdic acid or sodium molybate in 100 ml of hot concentrated sulfuric acid.
A.10 Mecke reagent	Dissolve 1.0 g of selenious acid in 100 ml of concentrated sulfuric acid.
A.11 Zwikker reagent	Solution A: Dissolve 0.5 g of copper (II) sulfate pentahydrate in 100 ml of distilled water. Solution B: Add 5 ml of pyridine to 95 ml of chloroform. Procedure: Add 1 volume of solution A to the drug, followed by 1 volume of solution B.
A.12 Simon's reagent	Solution A: Dissolve 1 g of sodium nitroprusside in 50 ml of distilled water and add 2 ml of acetaldehyde to the solution with thorough mixing. Solution B: 2% sodium carbonate in distilled water. Procedure: Add 1 volume of solution A to the drug, followed by 2 volumes of solution B.

Source: Table reproduced from Color Test Reagents/Kits for Preliminary Identification of Drugs of Abuse, NIJ Standard–0604.01, U.S. Department of Justice.

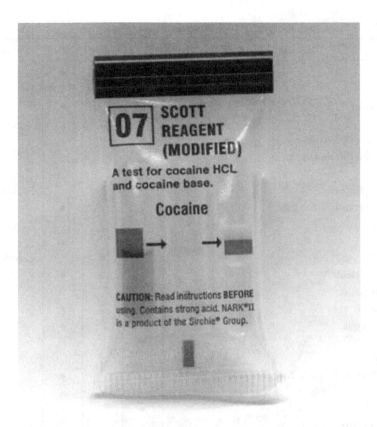

Figure 3.11 An example of a field cocaine test kit. This test uses cobalt thiocyanate to detect cocaine in acidic and basic environments.

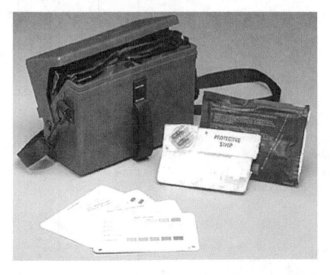

Figure 3.12 The M256A1 Kit includes instructions, M8 paper, and several vapor samplers to detect and identify chemical warfare agents as vapors in air.

Table 3.6 M8 Paper Response to Common Materials

Material	Neg	G	V	H	Comments
Acetone				•	
Acetonitrile	•				
Acrylic sealer				•	
Aloe gel	•				
Ammonium hydroxide		•			
Ammonium nitrate (aqueous)	•				
Antifreeze		•			
Arsenic trichloride					Brilliant violet
Automatic transmission fluid				•	
Bleach (household)	•				
Blood				•	
Brake fluid		•	•		Colors mixed
Carburetor cleaner				•	
Charcoal lighter	•				
Chloroform				•	
Clear nail polish				•	
Corn oil		•			
Corn syrup	•				
DDVP (Vapona®), 1% in Xylene				•	
Diazinon (household)				•	
Diesel fuel		•			
Dimethyl methylphosphonate		•			
Dimethylsulfoxide			•		
Dishwashing detergent	•				
Ethanol	•				
Ethylene glycol	•				
Gasoline				•	
Gear lube				•	
Glycerin	•				
Hexane	•				
Honey	•				
Hydrochloric acid, 3N	•				
Hydrogen peroxide (3% aqueous)	•				
Isopropyl alcohol	•				
Liquid Wrench®		•			
Lysol® Cling gel bowl cleaner (acidic)	•				
Maple syrup		•			
Methanol	•				
Methylene chloride				•	
Methyl ethyl ketone				•	
Mineral spirits	•				

(Continued)

Table 3.6 M8 Paper Response to Common Materials

Material	Neg	G	V	H	Comments
Motor oil	•				
Nitric acid (38% aqueous)	•				
Nonanone			•		
OFF! ® bug repellant (DEET)		•			
Olive oil		•			
Power steering fluid		•			
Sevin® (aqueous emulsion)	•				
Silicon lubricant	•				
Sodium hydroxide, 1N		•		•	Colors mixed
Sodium hydroxide (concentrated)		•			
Starter fluid (diethyl ether)	•				
Sulfuric acid (2N aqueous)	•				
Sulfuric acid (93% aqueous)	•				
Triethanolamine			•		
Triphenylphosphine (aqueous)	•				
Triphenylphosphine (hexane)	•				
Turpentine	•				
Vanilla oil		•			
Vinegar	•				
Water	•				
WD-40®	•				
Worcestershire sauce	•				
Windshield washer fluid (methanol)	•				
Xylene				•	

Source: Houghtons, Inc. and Cerberus and Associates, Inc., April 18, 2003.

harmful. M8 paper is intended to warn of the possibility of chemical warfare agent contamination. It is a presumptive test and produces many false-positives when used in nonbattlefield conditions (Table 3.6).

M8 paper is much less reliable in urban settings where man-made compounds are found and does not respond as readily to chemical agents if the paper is wet, including CW agents in water. M8 paper is helpful in determining if a liquid is organic or aqueous (inorganic) regardless of a color change. Organic liquids absorb into M8 paper while aqueous solutions bead up (do not absorb) on the M8 paper. Although simplistic, this method may contribute to identifying the presence of CWAs as they are all organic compounds, with some CWAs being soluble in water.

M8 paper contains carcinogenic compounds that produce a color change on light tan paper when the paper is dipped or wiped on a contaminated surface. The unbleached paper contains dyes that are released by exposure to some chemicals. G-nerve agents produce a yellow-brown, V-nerve agents produce a dark green, and H-blister agents produce a red-pink color.

Significantly, some commercial insecticides, which could be used as the basis for an improvised nerve agent, are not detected by M8 paper.

In addition to M8 paper, the M256A1 kit includes a colorimetric test on a card to detect and identify chemical warfare agents as vapors in air. The test can operate from −25° to 120°F (−32° to 49°C). Each card contains a test for blister agents, blood agents, nerve agents, and lewisite. Each test is run simultaneously, with results occurring over 20–25 minutes, which includes a 10-minute sampling time. False-negatives are unlikely, but false-positives can occur from common materials found in urban settings. The directions are straightforward, but lengthy. The instructions require one person to read the instructions and another to perform the steps. The test requires no additional support such as electric power, a reader, etc. Each test card contains 2.9 mg of mercuric cyanide and is subject to waste restrictions.

Chemical warfare agents detected are:

- Hydrogen cyanide (AC)
- Cyanogen chloride (CK)
- Mustard (H)
- Distilled mustard (HD)
- Nitrogen mustard (HN)
- Phosgene oxime (CX)
- Lewisite (L)
- V- and G-nerve agents and, presumably, A-series agents

A-series nerve agents, also know as novichoks or fourth-generation nerve agents, are cholinesterase inhibitors. A-series nerve agents share chemical similarities with both V- and G-nerve agents. A-series nerve agents do not yet appear in publicly available information for the M256A1 Kit following recent declassification. Compare the nerve agent test in the M256A1 Kit with the Agri-Screen Ticket® pesticide test in the following section, Cholinesterase Inhibitor.

Instructions for Use of the M256A1 Kit (from U.S. Army Technical Manual 3-6665-307-10)

1. Perform before operation Preventative Maintenance Checks and Services (PMCS) in accordance with TM 3-6665-307-10.
2. Prepare the kit for use.
   a. Read all the instruction cards in the kit.
   b. Remove one sampler detector from the kit and read the instructions printed on the bag.

Note: Do not expose sampler detector to heavy rain or other forms of water. Test results could be tainted.

Note: Do not touch sampler detector test spots. Dirt and oil from your gloves will cause test results to be tainted.

    c. Remove sampler detector from bag (save bag and use instructions printed on the outside). Dispose the sampler detector if there are broken or missing ampoules, missing spots, crushed reagent channels, or if the blood agent test spot is pinkish.

3. Test for toxic agent vapors.

    a. Swing out the heater; remove and save the two heater pads (used for breaking glass ampoules). Swing heater back in. Keep the protective strips over the spots.

    b. Remove pull tab (marked 1) to expose lewisite detecting tablet. Bend tab (marked 2) over lewisite detecting tablet and rub upper half of tab until a mark is visible.

Warning: Before Breaking Glass Ampoules (Except Heater Ampoules) Place One Heater Pad On Each Side Of Sampler Detector Covering The Ampoule To Be Broken. These Pads Will Prevent Pieces Of Glass From Cutting Your Gloves And Hands

    c. Hold sampler detector with test spots and arrow pointing up; crush four center ampoules (marked 3).

    d. Turn sampler detector so arrow points down. Using heater pads, squeeze ampoules to force liquid through formed channels into the test spots.

    e. Hold the sampler detector with arrow down and your thumb on the protective strip over the middle test spot.

      (1) Swing the heater away from the blister test spot.

      (2) Activate first heater ampoule (marked 4) by crushing one green ampoule, swing heater back over test spot, and leave in place for two minutes. Hold sampler detector to one side to avoid the vapor. DO NOT USE HEATER PADS TO CRUSH GREEN AMPOULES.

      (3) Swing heater and protective strip away from test spots.

Note: Do not hold sampler detector in direct sunlight while exposing the test spots. Test results could be tainted.

    f. Expose the test spots to air (shield from direct sunlight) for ten minutes. The sampler detector can be laid down or held by the hinged protective strip.

    g. After ten minutes, crush the second green ampoule (marked 4) and immediately swing heater over blister test spot. DO NOT USE HEATER PADS TO CRUSH GREEN AMPOULES. After one minute, swing heater away from test spot.

    h. Hold sampler detector with arrow pointing down and test spots exposed. Using heater pads, crush remaining ampoules (marked 5). Be sure to wet test spots by squeezing ampoules to force liquid onto them.

    i. Bend tab (marked 2) over lewisite detecting tablet and rub bottom half of tab until a mark is visible.

j. Turn the sampler detector upside down and compare colors
    of test spots (including lewisite tab) with those shown on
    sampler detector.
    (1) Compare blood agent (round spot) test after about ten
         minutes' exposure time.
         (a) Yellow or orange sometimes occur when no agent is
              present.
         (b) Pink or blue must be present to indicate blood agents.
              Any combination of colors, or rainbow effect, which
              includes pink or blue, should be considered a positive
              blood agent test.
    (2) Compare lewisite test after about ten minutes of expo-
         sure time. Look very closely at low concentrations; the
         color change may be very slight.
    (3) Blister agents (H and CX) develop color immediately
         after all ampoules are broken.
    (4) Nerve agent requires about a three-minute wait.
         (a) M256: If a peach color develops, a positive nerve test is
              indicated.
         (b) M256A1: If no color develops, a positive nerve test is
              indicated.
    (5) Disregard any small blue or blue-green areas under plas-
         tic rim of nerve agent spot.
k. Report the results to your supervisor.
l. Disposal instructions.
    (1) Dispose of expended or unserviceable materials IAW:
         federal, state, and local laws; military regulations and
         publications; host nation laws (if more restrictive than
         U.S. laws); and local Standard Operating Procedures. As
         a minimum, place used decontaminating materials in
         a sealed plastic bag (example: Ziploc bag), and label the
         bag with its contents.
    (2) Dispose of contaminated hazardous waste materials in
         accordance with FM 3-5 NBC Decontamination.
4. Perform after operation PMCS IAW TM 3-6665-307-10.

Cholinesterase Inhibitor

Cholinesterase inhibiting material is included in many compounds such as
military nerve agents, banned insecticides, registered pesticides, and some
medicines. Due to the potential toxicity of these materials, their unregulated
use poses a hazard. Cholinesterase inhibiting material has many chemical
formulations, which makes chemical identification of all these materials
difficult.

A chemical test for any material that inhibits cholinesterase becomes extremely useful in identifying the hazard and not necessarily identifying the substance. One such patented test is marketed under the name Agri-Screen Ticket. This is the same test used in the military M-272 Water Test Kit. The test uses stabilized electric eel cholinesterase on a pad. A sample suspected of containing cholinesterase inhibiting material is applied to the pad and allowed to bond to the eel cholinesterase, which chemically blocks the enzyme.

Cholinesterase is also capable of hydrolyzing indoxyl esters (usually indoxyl acetate) to a fluorescent indoxyl derivative, which is then oxidized to a colored species. This allows functioning cholinesterase to be visually detected by a color change. The indoxyl ester, referred to as the "developer," is in a second pad on the test ticket. The ticket is folded so that the pads maintain contact for three minutes. The instructions indicate the ticket should be held between the fingers so that the reaction is maintained at body temperature. If wearing thick gloves or in cold temperatures, an alternative is to clamp the ticket inside a folded hand warmer.

If the cholinesterase on the ticket is able to function normally, a white to blue color change will occur. If cholinesterase inhibiting material is present, the color change will be inhibited; a white pad indicates the presence of cholinesterase inhibiting material. Detection limits are displayed in Table 3.7.

This test must be kept dry until use and must be protected from hot temperatures. The test uses biochemicals, which are sensitive to degradation by heat. The manufacturer places a two-year expiration date on the test when kept at room temperature. It is estimated the test would degrade in a week if stored in a vehicle without air conditioning in the summer.

To assure the test is functional, you can modify it by cutting it and using one side of the pad as a control. As shown in Figure 3.13, a wedge shape is removed from the pad. Only one side of the white pad is exposed to the sample, the other to water. After a minute, the ticket is folded and held for three minutes. When opened, the control should show a color change if the test is functional. The other side will show the appropriate test result.

Air Monitoring Colorimetric Tubes

Air monitoring colorimetric tubes are tests sealed in glass units containing colorimetric reagents on an inert base, such as silica powder. Just before use, the operator snaps off the end of the tube and inserts the tube into a device that draws a set volume of air through the tube. A colorimetric reaction is produced by a target substance. The colorimetric reaction will proceed down the length of the tube as more air sample is drawn in. A scale on the tube is calibrated to the volume of air being pumped through and a quantitative result is possible. For this reason, these colorimetric air monitoring tubes are sometimes referred to as "length of stain" tubes.

Table 3.7   Agri-Screen Ticket Detection Limits

	Insecticides	
Common Names	Trade Names	Detection Limit in Water (ppm)
Carbamates		
Aldicarb	Temik®	0.2
Carbaryl	Sevin	7.0
Carbofuran	Furadan®	0.1
Mesurol	Methiocarb	5.0
Methomyl	Lannate®	1.0
MIPC	Isoprocarb	2.0
Oxamyl	Vydate L®	1.0
Propoxur	Baygon®	1.0
Organophosphates		
DDVP	Vapona	3.0
Methamidophos	Monitor®	4.0
Mevinphos	Phosdrin®	2.0
*Thiophosphates**		
Aspon		5.0
Azinphos-methyl	Guthion®	0.3
Chlorpyrifos-ethyl	Dursban®, Lorsban®	0.7
Chlorpyrifos-methyl	Reldan®	1.0
Diazinon	Spectracide®, Dianon®	2.0
EPN		0.2
Fenitrothion		1.5
Malathion		2.0
Metasystox-R		20.0
Methyl parathion		4.0
Parathion		2.0
Phorate	Thime®	3.0
Phosmet	Prolate®	1.0
Phosvel	Leptophos	0.8

*Requires conversion to oxygen analog for detection.

Note: These detection limits are based on a one-minute exposure of the Agri-Screen Ticket in an aqueous insecticide solution. Other techniques will lower detection limits.

Source: Reproduced by permission of Neogen Corporation, Lansing, MI, Copyright 2000.

Tests are designed to detect certain substances in controlled settings. There may be more than one colorimetric test for a target substance depending on the sampling range. These tubes are designed, for the most part, for use in an environment where the contaminant is known and a quantitative result is necessary. The tests may also be used in qualitative analysis, but

Figure 3.13 The Agri-Screen Ticket indication of the presence of a cholinesterase-inhibiting substance results from no color change on the ticket. A major complication is the complete failure of the test reagents, which also results in no color change. Test confidence is increased greatly if the ticket is modified as shown. A small wedge shape is cut from the test pad. Diluted suspect material is placed on one side and the other side is wetted with water. Do not allow the sample to touch the control side. Both sides are allowed to soak for 1 minute, and then the ticket is folded and held together for three minutes. The water-soaked side will act as a control.

a thorough knowledge of the reaction principle is needed in order to prevent confusion and inaccurate results. For example, Drager produces a tube called "Hydrochloric Acid 50/a" and its quantitative scale is set to 500–5000 ppm hydrochloric acid after drawing 1 L of air sample through the tube (Figure 3.14). The reaction principle is based on bromophenol blue, a pH indicator that changes from yellow at pH 3.0 to purple at pH 4.6; this reaction is reversible. The tube is intended for accurate quantitative measurement of hydrochloric acid in air when the contaminant is known. If you are familiar with the reaction principle and do not rely on the name of the test, you could use this test for the qualitative determination of any mineral acid in air and disregard the quantitative reading. In other words, this tube would function as a general acid in air detector and would be more sensitive than waving a piece of wet pH paper through the air.

Several colorimetric air monitoring tests are listed in Table 3.8 that might be useful when identifying hazardous characteristics of unknown material along with some practical comments. These tubes are intended for atmospheric monitoring; the presence of a liquid or solid sample may preclude the need for many of these air tests. Drager brand tubes are reviewed in Table 3.8. Other brands are available that use other tests with various results; no endorsement of any particular product is made here. Colorimetric air monitoring tubes are available from several manufacturers such as Drager, Gastec®, Sensidyne, and others.

Understanding the reaction principle for any one colorimetric tube is important when characterizing the hazards of an unknown gas. For example, if a Hydrocyanic Acid 2/a (Drager) tube is selected randomly for testing an unknown sample, a positive result does not necessarily mean hydrocyanic acid is present in the sample, but it certainly gives you some good leads. The tube is cross-sensitive to phosphine and high concentrations of other gases

Figure 3.14 Colorimetric tube in the hand-held Drager Accuro® pump (top). Drager ammonia tube detecting ammonium hydroxide vapor (bottom). Colorimetric tubes may be used qualitatively with a bulb syringe or other inexpensive pump.

may interfere. Using this tube as one, but not the only, source of information is helpful in the analysis.

Drager provides a combination of tests for qualitative analysis. Batches of tests are performed simultaneously and the results are compared to a matrix. This approach improves confidence in results because a finding is confirmed by a combination of resulting colors in more than one test that rule in or rule out certain gases. Drager's qualitative analysis scheme is summarized in Table 3.9. The test sets currently available are:

- Simultaneous Test Set I for inorganic fume: Developed for semi-quantitiative analysis of inorganic gases and vapors after a fire
- Simultaneous Test Set II for inorganic fume: Developed for semi-quantitiative analysis of inorganic gases and vapors after a fire
- Simultaneous Test Set III for organic vapors: Developed for semi-quantitative analysis of organic vapors

Table 3.8 Individual Colorimetric Air Monitor Tests for Qualitative Analysis

Test Name	Gas Test	Test Method	Comments
Acetaldehyde 100/a (Drager)	Aldehydes, ethers, ketones, petroleum hydrocarbons, and esters	Aldehydes convert chromium VI (yellow-orange) to chromium III (green).	A broad screen. The most practical detection of these compounds at higher concentration will be made with a combustible gas indicator. More complex molecules give a reduced sensitivity. Some aldehydes, ethers, ketones, petroleum hydrocarbons, and esters are indicated.
Acetic acid 5/a (Drager)	Acid	pH indicator.	This is a sensitive pH test in air. All acid gases produce a yellow color from blue. Compare to basic gas tube.
Acetone 100/b (Drager)	Ketones and aldehydes	Ketone + 2,4-dinitrophenylhydrazine → yellow hydrazone.	Useful as a screen for ketones and aldehydes.
Acid test (Drager)	Acid	Contains a pH indicator.	Cannot discern individual acids.
Acrylonitrile 0.5/a (Drager)	Nitriles	Chromium VI generates HCN from nitriles. HCN then reacts with HgCl$_2$ to form hydrochloric acid, which produces a red color with methyl red.	This method should be adaptable to other nitriles.
Alcohol 100/a	Alcohols	Alcohols convert chromium VI (yellow-orange) to chromium III (green).	A broad screen. The most practical detection of these compounds at higher concentration will be with a combustible gas sensor. More complex alcohols give a reduced sensitivity. Some aldehydes, ethers, ketones and esters are indicated.
Amine test (Drager)	Basic gases	pH indicator.	Amines can be detected as basic gases.
Ammonia 2/a (Drager)	Basic gases	pH indicator.	Amines can be detected as basic gases.

(Continued)

Table 3.8 (Continued) Individual Colorimetric Air Monitor Tests for Qualitative Analysis

Test Name	Gas Test	Test Method	Comments
Aniline 0.5/a (Drager)	Aniline	Aniline reduces orange chromium VI to green chromium III.	Sensitive to aniline and methyl aniline. Some ethers, ketones, aromatics, petroleum hydrocarbons, and esters are indicated.
Arsine 0.05/a (Drager)	Arsine and phosphine	Arsine and phosphine form a gray-violet color (colloidal gold) from Au^{+3}.	Sensitive test for arsine (0.05 ppm) and phosphine (3 ppm).
Benzene 0.5/a (Drager)	Aromatics	Two aromatic molecules are linked by formaldehyde in a condensation reaction. The linked aromatics are attacked by sulfuric acid and a pale brown color is produced.	A screen for many aromatics. Benzene as well as toluene, ethyl benzene, toluene and others are detected.
Carbon dioxide 5%/A (Drager)	Carbon dioxide	$CO_2 + N_2H_4 \rightarrow NH_2\text{-}NH\text{-}COOH$ (purple).	For the measurement of carbon dioxide in % scale. A conventional "multi-gas" air monitor will be more practical monitoring depressed oxygen level. This tube may be used to verify CO_2 in the case of other inert oxygen displacing gases.
Carbon tetrachloride 5/c (Drager)	Chlorinated hydrocarbons	$CCl_4 + H_2S_2O_7 \rightarrow COCl_2COCl_2 +$ aromatic amine $\rightarrow$ blue-green reaction product.	Test is not specific to carbon tetrachloride. Does not indicate all chlorinated hydrocarbons.
Chlorine 0.2/a (Drager)	Chlorine, bromine, and nitrogen dioxide	Chlorine gas and o-tolidine produce an orange reaction product.	Cross-sensitive to bromine, chlorine dioxide, and nitrogen dioxide. May be sensitive to other strong oxidizers.
Ethylene 0.1/a (Drager)	C=C bonds	C=C compound + Pd Molybdate complex $\rightarrow$ blue reaction product.	Screens for material containing carbon-carbon double bonds. Test does not discern material.

(Continued)

Table 3.8 (Continued) Individual Colorimetric Air Monitor Tests for Qualitative Analysis

Test Name	Gas Test	Test Method	Comments
Halogenated hydrocarbon 100/a (Drager)	Halogenated hydrocarbons	Tube contains a heat source that will pyrolize halogenated hydrocarbons. Chlorine radical forms HCl and the pH indicator produces yellow color from purple.	Test is an ignition source. Test only works while metal is hot. Detects all halogenated hydrocarbons. Cross-sensitive to halogens and hydrogen halides such as HF; HCl, HBr, and HI.
Hydrocyanic acid 2/a (Drager)	Hydrocyanic acid	$HgCl_2$ in tube produces HCl when HCN is present. Methyl red in tube produces red color with HCl	Cross-sensitive to phosphine. Do not use for HCl detection as 1000 ppm HCl does not interfere. High concentrations of other gases may interfere.
Hydrogen fluoride 1.5/b (Drager)	Hydrogen fluoride	HF reacts with $Zr(OH)_x$/quinalizarin complex to produce pale pink color.	Test is effective only for hydrogen fluoride in dry conditions. Moisture causes the formation of hydrofluoric acid, which cannot be indicated within the range of the test.
Nitrous fumes 0.5/a (Drager)	Nitrous gases	$NO + Cr^{VI} \rightarrow NO_2 NO_2 +$ diphenylbenzidine produce a blue-gray color.	Detects NO or NO_2. Greater than 300 ppm NO_2 will bleach the indication. Not effective in the presence of more than 1 ppm chlorine or 0.1 ppm ozone.
Phosgene 0.02/a (Drager)	Phosgene	Phosgene and an aromatic amine produce a red color	This test may be used to verify phosgene in the presence of chlorinated hydrocarbons. High concentration of phosgene will bleach the color indicator (>30 ppm). Chlorine and hydrochloric acid interfere.
Phosphine 0.1/a (Drager)	Phosphine, arsine, and antimony hydride	Phosphine and gold III forms a gray-violet colloid.	Test is sensitive for phosphine (0.1 ppm) but is cross-sensitive to arsine.

Table 3.9   Drager Colorimetric Test Sets for Qualitative Analysis

Test Set	Gas Tests	Test Method	Comments
Simultaneous test set I for inorganic fume	Acid gases	pH indicator.	This is a sensitive pH test in air. All acid gases produce a blue color from yellow. Compare to basic gas tube.
	Hydrocyanic acid	$HgCl_2$ in tube produces HCl when HCN is present. Methyl red in tube produces red color with HCl. Probably contains a filter to remove some interfering gases.	Cross-sensitive to phosphine. Do not use for HCl detection as 1000 ppm HCl does not interfere.
	Carbon monoxide	CO reacts with I_2O_5 and $H_2S_2O_7$ in tube to produce a brown iodine stain.	A conventional "multi-gas" air monitor will be more practical for CO monitoring. This tube may be used to verify CO in the case of gases that produce cross-sensitive results in the electronic monitor. This tube also detects acetylene, a Lewisite degradation product.
	Basic gases	pH indicator.	This is a sensitive pH test for air. All basic gases produce a blue color from yellow. Compare to acid gas tube and ammonia tubes.
	Nitrous gases	$NO + Cr^{VI} \rightarrow NO_2NO_2 + diphenylbenzidine$ produces a blue-gray color.	Detects NO or NO_2. Greater than 300 ppm NO_2 will bleach the indication. Not effective in the presence of more than 1 ppm chlorine or 0.1 ppm ozone.
Simultaneous test set II for inorganic fume	Sulfur dioxide	$SO_2 + I_2 + starch \rightarrow H_2SO_4$ and HI.	This is the complementary reaction to potassium iodide starch strips. SO_2 (and H_2S) are able to reduce I_2 (but not other halogens), which removes the blue color.
	Chlorine	Chlorine gas and o-tolidine produce an orange reaction product.	Cross-sensitive to bromine, chlorine dioxide and nitrogen dioxide.

(Continued)

Table 3.9 Drager Colorimetric Test Sets for Qualitative Analysis

Test Set	Gas Tests	Test Method	Comments
	Hydrogen sulfide	Production of mercuric sulfide to form a pale brown color.	A conventional "multi-gas" air monitor will be more practical for H_2S monitoring. This tube may be used to verify H_2S in the case of gases that produce cross-sensitive results in the electronic monitor.
	Carbon dioxide	$CO_2 + N_2H_4 \rightarrow NH_2\text{-}NH\text{-}COOH$ (purple).	For the measurement of carbon dioxide in % scale. A conventional "multi-gas" air monitor will be more practical monitoring depressed oxygen level. This tube may be used to verify CO_2 in the case of other inert oxygen displacing gases.
	Phosgene	Phosgene and an aromatic amine produce a red color.	This test may be used to verify phosgene in the presence of chlorinated hydrocarbons. High concentration of phosgene will bleach the color indicator (> 30 ppm). Chlorine and hydrochloric acid interfere.
Simultaneous test set III for organic vapors	Ketones	Probably a reaction of ketones and aldehydes with 2,4-dinitrophenylhydrazine to form various intensity of yellow hydrazone.	A broad screen. The most practical detection of these compounds at higher concentration will be with a combustible gas sensor.
	Aromatic hydrocarbons	Probably a reaction of the aromatic with I_2O_5 and $H_2SO_4 \rightarrow I_2$. There is a pretreatment area or filter in the tube. Compare to the aliphatic hydrocarbon tube below.	A broad screen. The most practical detection of these compounds at higher concentration will be with a combustible gas sensor.
	Alcohols	Alcohols convert chromium VI (yellow-orange) to chromium III (green).	A broad screen. The most practical detection of these compounds at higher concentration will be with a combustible gas sensor. More complex alcohols give a reduced sensitivity. Some aldehydes, ethers, ketones and esters are indicated.

(Continued)

Table 3.9 Drager Colorimetric Test Sets for Qualitative Analysis

Test Set	Gas Tests	Test Method	Comments
	Aliphatic hydrocarbons	Probably a reaction of most hydrocarbons with I_2O_5 and $H_2SO_4 \rightarrow I_2$.	A broad screen. The most practical detection of these compounds at higher concentration will be with a combustible gas sensor.
	Chlorinated hydrocarbons	Halogenated hydrocarbon forms chlorine gas after oxidation by permanganate. Diphenylbenzidine forms a blue color in the presence of chlorine gas and possibly other halogen gases.	A broad screen. Another method of detecting these compounds, even at low concentration, is a refrigerant gas detector.
CDS simultaneous test set I	Thioether (S-Mustard)	Gold chloride, thioether (R-S-R'), and chloramide forms an orange complex.	A screen for all thioethers. This test is not specific to sulfur mustard.
	Phosgene	Phosgene, ethylaniline and dimethylaminobenzaldehyde form a blue-green reaction product.	This method is different than the phosgene test in Test Set II. Low amounts of hydrochloric acid (<100 ppm) do not interfere. Test is cross-sensitive to acetyl chloride and carbonyl bromide.
	Hydrocyanic acid	$HgCl_2$ in tube produces HCl when HCN is present. Methyl red in tube produces red color with HCl.	Cross-sensitive to phosphine. Do not use for HCl detection as 1000 ppm HCl does not interfere. Same test as Set I.
	Organic arsenic compounds and arsine (Lewisite)	Step one collects a sample in the reaction media. Arsine separates gold from a complex to form a gray to violet color. In step two, zinc and hydrochloric acid reduce organic arsenic to arsine and with more pump strokes is indicated by the reaction from step one.	Step one is sensitive to arsine and phosphine. Step two detects organic arsenic (As^{+3}) compounds. Both results appear in the same colorimetric area of the tube. Arsine and phosphine can produce a gray color and colloidal gold can form rose, purple, or yellow color based on concentration. The test lists a measuring range of 0.1 ppm arsine and 3 mg/m^3 of organic arsenic.

(Continued)

Table 3.9 Drager Colorimetric Test Sets for Qualitative Analysis

Test Set	Gas Tests	Test Method	Comments
CDS simultaneous test set V	Organic basic nitrogen compounds (nit. mustard)	Organic amine + $KBiI_4 \rightarrow$ orange-red reaction product.	A general screening test for organic amines. Individual compounds cannot be identified.
	Cyanogen chloride	Cyanogen chloride and pyridine will form glutaconaldehyde cyanamide. Glutaconaldehyde in the presence of barbituric acid forms a pink reaction product.	Test is sensitive to 0.25 ppm cyanogen chloride. Cross-sensitive to cyanogen bromide. This test could be used to verify cyanogen chloride from other halogenated hydrocarbons.
	Thioether (S-mustard)	Gold chloride and thioether (R-S-F) forms a bright yellow to orange complex.	A screen for all thioethers. This test is not specific to sulfur mustard. Same as CDS Set I.
	Phosgene	Phosgene, ethylaniline, and dimethylaminobenzaldehyde form a blue-green reaction product.	This method is different than the phosgene test in Test Set II. Low amounts of hydrochloric acid (<100 ppm) do not interfere. Test is cross-sensitive to acetyl chloride and carbonyl bromide. Same test as CDS Set I.
	Chlorine	Chlorine gas and o-tolidine produce an orange reaction product.	Cross-sensitive to bromine, chlorine dioxide, and nitrogen dioxide. Same test is in Simultaneous Test Set II for inorganic fume.
	Phosphoric acid ester (nerve agents, e.g., Tabun, Sarin, and Soman)	If phosphoric acid ester is present, cholinesterase in the tube is inactivated after a sample is drawn into the tube and one-minute passes. Butylcholine iodide and water are added and if the cholinesterase is active, it will cleave butylcholine iodide and form butyric acid. Butyric acid and phenol red pH indicator produces a yellow color (negative result). If the cholinesterase has been inactivated, butyric acid will not form and the phenol red pH indicator will produce a red color (positive result).	Sensitive test for airborne phosphoric acid esters (diclorvos 0.05 ppm) but detects others at varying sensitivity. Reference does not clearly state if other cholinesterase inhibiting materials (carbamates, thiophosphates, and phosphonates) are detected at similar sensitivity.

(Continued)

Table 3.9 Drager Colorimetric Test Sets for Qualitative Analysis

Test Set	Gas Tests	Test Method	Comments
Clan lab simultaneous test set	Phosphine	Phosphine and gold III forms a gray-violet colloid.	Test is sensitive for phosphine (0.1 ppm) but is cross-sensitive to arsine.
	Phosgene	Phosgene and an aromatic amine produce a red color.	This test may be used to verify phosgene in the presence of chlorinated hydrocarbons. High concentration of phosgene will bleach the color indicator (> 30 ppm). Chlorine and hydrochloric acid interfere.
	Ammonia	pH indicator.	This is a sensitive pH test for air. All basic gases produce a yellow to blue color. Compare to acid gas tube.
	Hydrochloric acid	Bromophenol blue pH indicator and hydrochloric acid produces a yellow color.	This is a sensitive pH test for air. All acid gases produce a blue to yellow color. Compare to acid gas tube in Set I. Hydrochloric acid cannot be discerned from other mineral acids.
	Iodine	Yellow changes to pale pink in the presence of iodine. Reaction principle not available.	

- CDS Simultaneous Test Set I: Developed for semiquantitative analysis of eight chemical warfare agents
- CDS Simultaneous Test Set V: Developed for semiquantitative analysis of eight chemical warfare agents
- Clan Lab Simultaneous Test Set: Designed for identification of chemical substances commonly associated with production of methamphetamine

These tests are intended for use when certain criteria are met, such as semiquantitative measurement of combustion gases after a fire. When using outside test parameters, you must understand the reaction principles in order to accurately characterize the hazards of an unknown gas. A negative result does not preclude other harmful gases.

Sensidyne Inc. produces a qualitative screening test for gases called the Deluxe Haz Mat III Kit (Figure 3.15). Three tubes containing several tests ("sections") are used in a prescribed method and the results are compared to a matrix. The first tube detects mostly inorganic gases. The second tube detects mostly organic gases. The third tube is identical to the second tube, but the flow of air is reversed through the sections. In the second tube, the air-flow is from sections A to D; in the third tube, airflow is from sections D to A. The initial three tubes are followed by individual tubes if indicated. There are five individual tubes indicated if the initial three tubes do not produce a result.

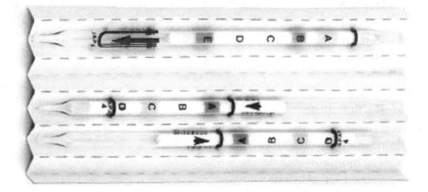

Figure 3.15 Sensidyne Tube 131 for inorganic vapors (top) and a pair of Sensidyne Tube 186B for organic vapor (bottom). Each section reacts to certain gases or vapors and provides filtering for subsequent layers.

Sensidyne Tube 131 (inorganic vapors) will identify the following compounds or families:

- Acetic acid
- Acetylene
- Amines
- Ammonia
- Arsine
- Carbon monoxide
- Chlorine
- Chlorine dioxide
- Hydrazine
- Hydrogen chloride
- Hydrogen sulfide
- Mercaptans
- Nitrogen dioxide
- Phosphine
- Sulfur dioxide

Testing with a pair of Sensidyne Tube 186B (organic vapor) will identify the following compounds or families:

- Aliphatic hydrocarbons
  - Propane
  - Butane
  - Hexane
  - Pentane
  - Kerosene
  - Heptane
- Alkenes
  - Ethylene
  - 1,3-Butadiene
- Acetylene
- Halogenated hydrocarbons
  - Trichloroethylene
  - Tetrachloroethylene
  - Vinyl Chloride
  - 1,1,1-Trichloroethylene
- Aromatic hydrocarbons
  - Gasoline
  - Benzene
  - Toluene
  - Xylene (all isomers)
  - Ethylbenzene

- Phenol
- Cresol (all isomers)
- Styrene
- Ketones
  - Acetone
  - Methyl ethyl ketone (MEK)
  - Methyl isobutyl ketone (MIBK)
- Acetates
  - Ethyl acetate
  - Butyl acetate
- Alcohols
  - Methanol
  - Ethanol
  - 1-Butanol
  - Isopropyl alcohol
- Aldehydes
  - Acetaldehyde
  - Formaldehyde
- Ethyl cellosolve (2-ethoxyethanol)
- Ethylene oxide
- Tetrahydrofuran
- Carbon monoxide
- Carbon disulfide
- Mercaptans
  - Ethyl mercaptan
  - Methyl mercaptan
  - Tert-butyl mercaptan
- Hydrogen sulfide
- Arsine
- Aniline
- Ethyl amine

Negative results indicate the need to test with five individual tubes to detect gases not determined in the initial screen.

The screen is more than a compilation of individual gas detection tubes. By selecting individual colorimetric tests for compounds when more than one is available begins to define parameters of the screen. The physical arrangement of test sections in different tubes and the order in which the sample is filtered through the tests also defines parameters. The screen is meant to be qualitative with additional testing as required for a few specific gases and vapors that do not produce a color change in the initial tests. As with any test, minimum detection limits may not be low enough to detect all gases present. Coloration produced outside the matrix results is not conclusive. Table 3.10 summarizes the Sensidyne qualitative analysis scheme using

Table 3.10 Sensidyne Colorimetric Test Sets for Qualitative Analysis

Test Set	Gas Tests	Test Method	Comments
Inorganic Gas Qualitative Detector Tube (#131), Section A	Basic gases	pH indicator.	This is a sensitive pH test in air. All basic gases produce a yellow color from light purple.
Section B	Acid gases in general with characteristic changes for hydrogen chloride and chlorine	pH indicator.	This is a sensitive pH test in air. All acid gases produce a yellow color from red-purple. Hydrogen chloride produces pink from red-purple. Chlorine bleaches the color to white.
Section C	Nitrogen dioxide and chlorine	Dye production from reaction with o-toluidine.	Nitrogen dioxide and o-toluidine form a dye, nitroso-o-toluidine, which is indicated as yellow. Reaction with chlorine produces a more orange yellow.
Section D	Hydrogen sulfide	Reaction with lead acetate.	Lead sulfide is visible as a brown color at low concentrations. This is the same reaction used in lead acetate test strips for sulfides in water.
Section E	Carbon monoxide, phosphine, acetylene, and methyl mercaptan	A palladium compound is reduced to free palladium.	Palladium produces different colors dependent on the gas present. Carbon monoxide produces a pale dark brown. Phosphine produces black. Acetylene produces a pale yellow-green. Methyl mercaptan produces dark yellow.

(Continued)

Table 3.10 Sensidyne Colorimetric Test Sets for Qualitative Analysis

Test Set	Gas Tests	Test Method	Comments
Organic Gas Qualitative Detector Tube, "A" side (#186B), Section A. Note: In the first draw of this tube (airflow A to D), Section A is the only test result used.	Organic gases and vapors in general, arsine, carbon monoxide, and hydrogen sulfide	Chromium VI is reduced to green chromium III. Due to the presence of some compounds, the green reaction product may be visible as another color due to mixing. Dark brown is the most likely color, but light brown, orange and green are also possible. Each color directs the operator through the test result matrix.	A broad screen. The most practical detection of most of these compounds at higher concentration will be with a combustible gas sensor. More complex alcohols give a reduced sensitivity. Some aldehydes, ethers, ketones, and esters are indicated. Refer to the *Sensidyne Gas Detector Tube Handbook* for a list of detected organic gases and vapors. Inorganic compounds detected are arsine, carbon monoxide, and hydrogen sulfide.
Organic Gas Qualitative Detector Tube, "D" side (#186B), Section D. Note: In the second draw of this tube (airflow D to A), all sections are used as test results.	Phenol, cresol, aniline, and ethyl amine	Compounds react with Ce^{4+} to form polymerization products.	Phenol and cresol form a pale brown. Aniline forms a blue-green. Ethyl amine forms a pale blue.
Section C	Butadiene, gasoline, ethylene, acetylene, styrene, methyl mercaptan, hydrogen sumde, arsine, and carbon monoxide	Iodine pentoxide is reduced to form iodine. Specifically: Substance + $I_2O_5 + H_2SO_4 \rightarrow I_2$,	Butadiene reacts with the test reagents and bleaches the test section white from the original yellow. Gasoline, ethylene, acetylene, and carbon monoxide produce a pale blue. Styrene and methyl mercaptan produce a yellow-orange. Hydrogen sulfide and arsine produce black.

(Continued)

Table 3.10 Sensidyne Colorimetric Test Sets for Qualitative Analysis

Test Set	Gas Tests	Test Method	Comments
Section B	Benzene, toluene, ethyl benzene, and toluene (BTEX)	Molybdate is reduced to molybdenum blue from reaction with palladium sulfate and C=C.	Presumably C=C compounds have been screened by test sections C and D and only the BTEX compounds can pass to this section and react.
Section A	Organic gases and vapors in general, arsine, carbon monoxide, and hydrogen sulfide	Chromium VI is reduced to green chromium III. Due to the presence of some compounds and the flow of the sample gas through previous sections, the green reaction product will be visible as another color due to mixing. Dark brown is the most likely color, but light brown and orange are possible. Each color directs the operator through the test result matrix.	A broad screen. The most practical detection of most of these compounds at higher concentration will be with a combustible gas sensor. Due to the screening effect of the previous sections, results can differ somewhat from Section A in the first draw. Some aldehydes, ethers, ketones, and esters are indicated. Refer to the *Sensidyne Gas Detector Tube Handbook* for a list of detected organic gases and vapors. Inorganic compounds detected are arsine, carbon monoxide, and hydrogen sulfide.

Table 3.11 Subsequent Testing Required as Indicated by Sensidyne Haz Mat III Testing Chart

Gas Test*	Positive Result Indicates	Negative Result Indicates	Comments
Arsine (Tube#121U)	Mercaptans, phosphine, arsine, diborane, hydrogen selenide, and hydrogen cyanide produce a pink color.	Acetylene and carbon monoxide allow the original pale yellow color to remain.	Gases producing a pink result are distinctly different yet all are very toxic.
Trichloroethylene (Tube #134SB)	Trichloroethylene, tetrachloroethylene, dichloroethylene, and vinyl chloride produce a purple color.	Negative results should continue with Carbon Disulfide Tube #141SB.	The gases/vapors that produce a purple result are from heavier than water, water-insoluble liquids.
Carbon disulfide (Tube #141SB)	Carbon disulfide and carbonyl sulfide produce a yellow color.	Negative results indicate shorter aliphatic hydrocarbons.	The gases/vapors that produce a yellow result are from heavier than water, water-insoluble liquids. Negative results are from lighter than water liquids or are gases at STP.
Benzene (Tube#118SB)	Benzene produces a green-brown color in the analyzer tube.	Other aromatic hydrocarbons are indicated by no response in the analyzer tube or by discoloration in the pretube.	Benzene identification is significant due to its toxicity; however, other aromatic hydrocarbons are not identified.
Acetone (Tube #102SC)	Aldehydes and ketones are indicated by a pink result.	Continue with Ethyl Acetate Tube #111U.	Aldehydes and ketones are indicated in the screen as the only remaining compounds that can liberate HCl from hydroxyl amine hydrochloride.
Ethyl acetate (Tube #111U)	Alcohols, esters, acetates, ketones, kerosene, and longer aliphatic hydrocarbons produce a brown color.	Continue with Methyl Bromide Tube #157SB.	These vapors/gases are indicated in the screen as the only remaining compounds that can reduce chromium VI.

(Continued)

Table 3.11 (Continued)   Subsequent Testing Required as Indicated by Sensidyne Haz Mat III Testing Chart

Gas Test*	Positive Result Indicates	Negative Result Indicates	Comments
Methyl bromide (Tube #157SB)	Methyl bromide, trichloroethane, chloropicrin, chloroform, and dichloromethane produce a yellow color.		This group of vapors/gases is the most toxic subset. Additional testing is necessary to identify a particular gas; all can be dangerous.
Hydrogen sulfide (Tube #120SB)	Hydrogen sulfide produces a dark brown result.	Continue with Tube#121U. Phosphine/Arsine.	This test separates toxic hydrogen sulfide from toxic arsine and other less toxic gases.
Phosphine/Arsine (Tube#121U)	Arsine produces a pink color.	Carbon monoxide and acetylene.	Highly toxic arsine is identified from less toxic carbon monoxide and acetylene.
Additional tubes to be used if the first three multisection tubes are negative	Carbon dioxide (Tube #126SA) Hydrogen (Tube #137U) Hydrogen cyanide (Tube#112SB) Hydrogen fluoride (Tube #156S) Nitric oxide (Tube #174A) Acetic acid (Tube#216S) Carbon tetrachloride (Tube #147S) Methyl bromide (Tube #157SB)		These tests are used to identify gases that are present outside the qualitative or quantitative detection parameters of the screen.

*The individual gas tests are valid only in conjunction with the Sensidyne Deluxe Haz Mat III test scheme.

the three multi test tubes. Table 3.11 follows and describes individual tests if the broad screen is negative.

As described in Chapter 2 in the section on water-reactive hazards, DOT has defined 70 water-reactive toxic inhalation hazard (WRTIH) substances as those that emit one of 12 toxic gases. Performance of the Sensidyne Deluxe Haz Mat III qualitative analysis scheme based on the manufacturer's information is shown in Table 3.12.

The test scheme above either identifies or characterizes WRTIH gases. Chlorine and bromine are identified and not clearly separated, but both have similar hazards. In high concentration, chlorine will have a yellow to green appearance and bromine will have a yellow to red-brown appearance. Likewise, hydrogen chloride and hydrogen bromide are identified and not strongly separated, but both have similar hazards. No indication from the screening tube (131) requires additional testing for hydrogen fluoride and hydrogen cyanide. Hydrogen sulfide is uniquely identified by lead acetate. Ammonia could be confused with a fuming basic compound, and testing on the liquid is necessary. Phosphine could be confused with arsine, but arsine is not legally transportable. Sulfur dioxide can be reasonably identified but sulfur trioxide will demonstrate a strong acidic, but nonspecific, quality. Detection of sulfur by some other form of technology would be helpful in identifying sulfur trioxide.

Electrochemical Sensors and Multisensor Monitors

Multisensor monitors and other instruments that utilize electrochemical sensors may be used to characterize unknown substances (Figure 3.16).

"Multi-gas" monitors grew out of the need to test for fire hazard, which was originally detected with a single sensor combustible gas indicator (CGI). These monitors typically contain four sensors, one each for combustible gas, oxygen, carbon monoxide, and hydrogen sulfide. "Four gas" instruments may have interchangeable sensors to expand the number of gases that can be presumptively identified. Modern instruments may integrate a photoionization detector (PID) for ppm level detection of volatile organic compounds, or VOC (see the Photoionization section for more detail).

Flammable vapors in air are detected by the combustible gas sensor. This sensor contains two catalytic beads, only one of which is exposed to the sample atmosphere. Electric circuitry measures the temperature difference between the control bead and the sample bead. If the sample atmosphere contains anything that will burn, it is catalyzed on the bead and the temperature of the sample bead rises above the control bead. Any combustible gas in the air will produce a flammable reading. The sensor is calibrated to a control gas, so if the sample gas is known, a response factor may be applied for an accurate quantitative value.

Table 3.12 Sensidyne Deluxe Haz Mat III Detection of WRTIH Gases

Symbol	Chemical	Tube 131 Indication	Comment
Br_2	Bromine	Produced yellow in the B and C layers, and brown in the E layer.	This result is not in the ID chart but was provided by the manufacturer. Follow with Tube 114 (Br_2) and compare to chlorine.
Cl_2	Chlorine	Produced white discoloration in the B layer. Produced yellow in layer E.	From Unique Color ID Chart.
HBr	Hydrogen bromide	Expected to produce yellow (as acid) or pink (as halogen) in the B layer.	This result is not in the ID chart and is extrapolated. There is no hydrogen bromide tube to continue testing.
HCl	Hydrogen chloride	Produced pink in the B layer.	From Unique Color ID Chart. There is no reason to follow with Tube 173SA (HCl) because it is another pH indicator.
HCN	Hydrogen cyanide	Not indicated. Some concentrations of HCN produced a pale yellow discoloration in the A layer and a white discoloration in the E layer.	This result is not in the ID chart but was provided by the manufacturer. Follow both results with Tube 112SB (HCN).
HF	Hydrogen fluoride	Not indicated.	From Unique Color ID Chart. There is no reason to follow with Tube 156S (HF) because it is another pH indicator. If HF can be captured in a water sample, the fluoride spot test can be used to confirm HF.
H_2S	Hydrogen sulfide	Produced brown in the D layer.	From Unique Color ID Chart.
NH_3	Ammonia	Produced yellow in the A layer.	From Unique Color ID Chart. Any basic gas (amines, hydrazine) will produce this result.
PH_3	Phosphine	Produced gray to blue in the E layer.	From Unique Color ID Chart. Follow with Tube 121U (arsine) for other possibilities.
SO_2	Sulfur dioxide	Produced yellow in the B layer.	From Unique Color ID Chart. Acetic acid is possible. Follow with Tube 103SA (SO_2).
SO_3	Sulfur trioxide	SO_3 is expected to produce yellow (as acid) in the B layer.	This result is not in the ID chart and is extrapolated. There is a sulfuric tube to continue testing, but it is not reliable due to interference. Sulfur trioxide will immediately form sulfuric acid in water.

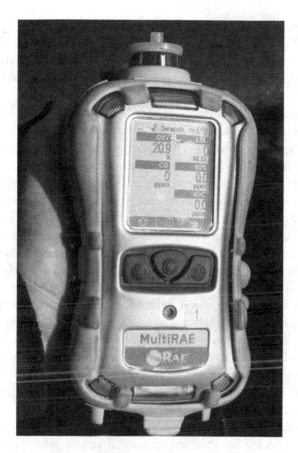

Figure 3.16 RAE multi-gas monitor with PID.

The other sensors in a multi-gas monitor utilize electrochemical sensors. Each electrochemical sensor contains a chemical substance that is reactive to the target compound. When the sample compound is absorbed into the sensor, the change in electrical conductivity, usually an oxidation reduction reaction, occurs, which generates a small electrical current or a change in conductivity. This reaction usually involves a catalyst embedded in the sensor. Three electrical leads measure changes in the electrical field within the sensor and pass the signal to a processor. The monitor displays the sample concentration in air based on calibration to a standard gas.

Oxygen is detected by an electrochemical sensor that uses a reducible polymer to measure the concentration of oxygen in air. It is referenced to normal atmospheric oxygen concentration of 20.9%. High oxygen, defined as greater than 23.5%, is a fire hazard in that it increases the combustibility of flammable vapors and other combustible substances. Low oxygen, defined as less than 19.5%, works to decrease fire hazard but obviously becomes an inhalation hazard.

The remaining two gas sensors were developed for confined space entry concerns. Carbon monoxide is the most common man-made gas hazard found in confined spaces, and hydrogen sulfide is the most common natural gas hazard found in confined spaces. Each of these gases is detected by using an electrochemical sensor that contains a chemical substrate sensitive to oxidation or reduction by the target gas. A signal is generated in a manner similar to the oxygen sensor.

The electrochemical sensors are rarely sensitive to only the target gas. Often the sensors are cross-sensitive to other gases. You must determine if the indicated gas is present or whether it might be a cross-sensitive gas. You can use cross-sensitive indications and situational awareness to characterize unknown substances. Some monitors utilize a chemical filter over the sensor in an effort to make the electrochemical sensor more specific to the target chemical. For example, some carbon monoxide sensors are cross-sensitive to isopropyl alcohol. In this case, you may see a combustible gas indication on the display at the same time the carbon monoxide result indicates a very high value.

Many other electrochemical sensors are available (Table 3.13). Most are cross-sensitive to other than the target chemical and are also sensitive to changes in temperature. Each detector manufacturer has the option of producing their own electrochemical sensors or using a sensor supplied by a vendor, so it is difficult to describe all the combinations of materials that can be detected. Understanding your monitor's cross-sensitivities will allow you to characterize an unknown chemical far beyond the quantitative results displayed by the monitor. Some electrochemical sensors are used in instruments outside the "multi-gas" configuration to form monitors dedicated to specific material detection. For example, cholinesterase inhibiting material in air can be detected by an electrochemical sensor. The sensor contains a known amount of cholinesterase that is exposed to the air. If a nerve agent is present, the cholinesterase is inhibited. Next, the sensor is exposed to a precise amount of reagent that is electrochemically reactive with intact cholinesterase. The resulting electrochemical signal is inversely proportional to the amount of cholinesterase inhibiting material present.

Table 3.13   Common Gases and Vapors Detectable by Electrochemical Sensors

Amines	Hydrogen fluoride
Ammonia	Hydrogen peroxide
Carbon dioxide	Hydrogen sulfide
Carbon monoxide	Nitric oxide
Chlorine	Nitrogen dioxide
Hydrazine	Organic vapor
Hydrides	Oxygen
Hydrogen	Phosphine
Hydrogen chloride	Sulfur dioxide
Hydrogen cyanide	

Another type of sensor that has been incorporated into a multiple sensor detector is the metal oxide semiconductor (MOS). These detectors measure small changes in temperature and electrical conductivity as the result of gas adsorption on the metal oxide or metal foil surface. These are mostly useful for measuring changes from normal background reading. This is a nonselective device because different contaminants have different thermal conductivities and therefore different responses.

Refrigerant Detector

Air monitors that detect refrigerants are a low cost and effective addition to any field characterization of unknown hazardous materials (Figure 3.17). Used for years by refrigeration technicians, these monitors detect part per million levels or less of halo hydrocarbon refrigerant gas. The capability allows the refrigerant detector to be used to detect nearly all halo hydrocarbons. Halo hydrocarbons are any hydrocarbon that contains a hydrocarbon structure with one or more covalently bonded fluorine, chlorine, bromine, or (sometimes) an iodine atom. Bromo and iodo hydrocarbons are a much less frequent problem than are the more commercially useful fluorinated and chlorinated hydrocarbons. Older monitors were used only to detect chlorinated hydrocarbons. The push for more environmentally friendly ozone introduced more fluorinated hydrocarbon refrigerants, which respond to the same detection techniques but are detected 20–100 times less sensitive than chlorinated hydrocarbons.

These devices function by detecting a change in gas levels and do not make a quantitative measurement. When powered on, the detector will "zero" itself to the background level of halo hydrocarbon at that moment. Some detectors will auto zero every ten seconds if the detector stops moving because the units are designed for a refrigeration technician to locate a leak in a system while being guided by an increasing tick rate from the monitor.

Figure 3.17 TIF 5660 refrigerant detector by TIF Instruments, Inc.

This technique of locating a leak by detecting an increase in concentration allows a technician to find the source of a leak without making manual sensitivity adjustments or being concerned about existing background vapors in the area. Other refrigerant detectors have a manual adjustment to compensate for increased concentration of halo hydrocarbon in air. Spinning the adjustment will decrease the tick rate; moving into a higher concentration will increase the tick rate. Other detectors may have a reset button that auto zeros to ambient concentration while others are simply turned off and back on with warm-up times of 10 seconds or less.

Refrigerant detectors are used to locate very slow leaks. These detectors are capable of locating a leak of Freon 134a (1,1,1,2-tetrafluoroethane, the most difficult to detect commercial refrigerant) of 0.5 ounce (14 g) per year. Detectors may use conductive polymer ionization, corona discharge, heated diode, or infrared methods of detection with sensitivities of 0.1–0.5 ounce per year. Field use does not typically require this level of sensitivity.

Refrigerant detectors for use in emergency characterization of hazardous material should utilize a disposable sensor that is immediately changeable. Some monitors have an additional filter to prevent introduction of dust and moisture.

When using the refrigerant sensor, note that exposing this sensor to a steady stream of highly concentrated refrigerant will severely reduce sensor life or damage the sensor. Sensor life is directly proportional to the amount of refrigerant that passes through the sensor.

False halo hydrocarbon indications are usually caused by abrupt changes in sensor temperature. These temperature changes are typically due to a sudden change in airflow past the sensor, sudden cooling, or the sensor being heated by an outside source. To avoid false readings, keep the sensor dry, shield it from wind while allowing continuous intake, and avoid sudden temperature changes. False indications can be induced by interfering gases and vapors, but the detector is less sensitive to nonhalogenated compounds. Use the detector to verify or locate halogenated materials, but do not assume that an initial response by a refrigerant detector immediately indicates a halogenated hydrocarbon.

Sensor overload may occur in areas of sudden, high concentration of halo hydrocarbon. Consult the detector manual for indication of overflow. If overflow occurs, the detector must be removed to clean air and allowed to stabilize or you can change the sensor tip.

Refrigerant sensors are designed to last about one year, assuming 150 hours per year of service. The sensor life is greatly diminished if exposed to high levels of halo hydrocarbon.

Ethylene oxide, a sterilization gas used in hospitals, sometimes contains a halo hydrocarbon to eliminate its explosive hazard. A refrigerant detector may be used to verify a release of ethylene oxide, which contains a halo hydrocarbon (DOT label for this material is green, compressed gas). The detector can also detect dry cleaning solvent (perchloroethylene) and Halon fire extinguishing agent in very low concentrations. SF-6, sulfur hexafluoride, is an insulating gas found in high voltage equipment that can be detected with a refrigerant detector.

Refrigerant detectors are not intended for the detection of chlorine and fluorine containing chemical warfare agents since the detection threshold is above IDLH.

Examples of refrigerant detectors are TIF XP-1A, Bacharach the Informant 2 Leak Detector, TPI 750a, and TIF 5750a Super Scanner Halogen Leak Detector. The Inficon D-TEK™ Select Refrigerant Leak Detector uses an infrared cell to improve sensitivity to fluoro hydrocarbons and reduce false alarms from smoke and moisture. However, it does not detect R-11 or R-123 and may be limited on other halo hydrocarbon compounds. Refrigerant detectors are inexpensive and easy to operate and maintain.

Electron Capture Device

The electron capture device (ECD) is used to detect electrophilic (electron attracting) molecules. It is especially sensitive to organic compounds containing chlorine or bromine; organometallic compounds, nitriles, and nitro compounds.

The ECD contains a metal foil with a small amount of radioactive nickel-63, which emits a steady stream of electrons and sets up a background current. As electrophilic compounds enter the current, electrons are absorbed and the corresponding drop in current is detected. When considered with retention time through a gas chromatograph, quantitative measurement and identification of the sample vapor is possible.

The ECD is up to 1,000 times more sensitive than a flame ionization detector, but is limited in practical use to mainly halogenated hydrocarbons. Since oxygen is an electrophile, ECD must be used in conjunction with a gas chromatograph and an inert carrier gas such as pure nitrogen. The ECD cannot be used as an air monitor.

Thermometer

An infrared (IR) thermometer measures temperature from a distance by detecting the amount of thermal electromagnetic radiation emitted from the object (Figure 3.18). This allows temperature sensing without contacting the sample material or its container.

The IR thermometer measures the surface temperature, not the core temperature, of an object. The unit's optics sense emitted (E), reflected (R), and transmitted (T) IR energy, which is focused onto the sensor. The resulting

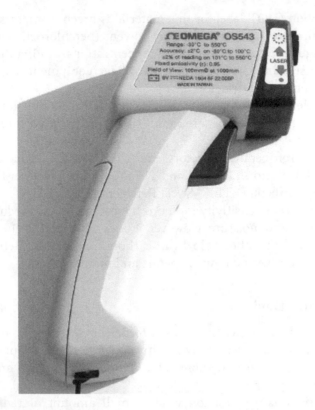

Figure 3.18 An example of an infrared thermometer with laser sighting, the OS543 from Omega Engineering, Inc.

electronic signal is processed and compared to the temperature of the detector at the time of the reading and is displayed as a temperature (Figure 3.19). Some IR thermometers have a laser that is used for pointing the sensor at the surface of the object. The laser does not affect the result.

Each model of IR thermometer will have a specified distance and spot (D:S) ratio. This indicates the sensor field of view. The thermometer will "see" everything within the field of view, so care should be taken to ensure the target being measured fills the field of view. As the distance from the target increases, the spot to be measured grows larger. The most accurate readings are those in which the IR thermometer is close to the target. For an accurate reading, make sure the target is at least twice the size of the spot.

Smoke or dust in the air can interfere with the reading by obstructing the IR energy as received by the thermometer. The IR thermometer cannot sense through a visually clear object such as glass; it will instead measure the temperature of the surface of the clear object.

Emissivity is the term that describes the ratio of IR energy radiated by an object compared to the IR energy of a standard. The IR thermometer works

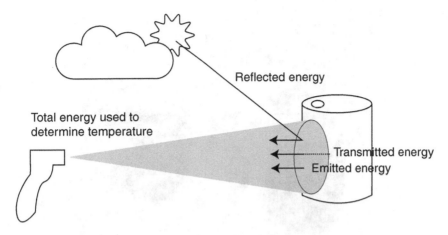

Figure 3.19 Infrared energy in total is used by the detector to determine temperature. A combination of reflected, emitted, and transmitted energy affects the infrared sensor. The sensor will collect information from areas within the "spot" that should completely cover the area to be measured.

by measuring emissivity. When the energy spectrum from an object is close to the thermometer's standard, the difference is more difficult to measure and inaccurate readings are possible. Some IR thermometers have a preset emissivity that may be very close to organic material, polished surfaces, or oxidized surfaces. To compensate for a surface that may have an emissivity of the thermometer's preset value, simply cover the spot with black tape, allow it time to assume the temperature of the object, and measure the new surface.

The scientific name for an IR thermometer is "radiation pyrometer." Other common names are "noncontact thermometer" and "laser thermometer" if the unit has an aiming laser. Pointing lasers are an eye hazard and care should be taken not to point them directly into a person's eye or onto reflective surfaces.

These devices are generally not protected from radio frequency interference and should be tested against portable radios. The units should be protected from exposure to arc welders, static electricity, etc. Since the IR thermometer compares the observed IR energy to the temperature of the detector unit, care should be taken to protect the IR thermometer from large changes in temperature while taking measurements. If used in excessively warm or cold environments, temperature readings may be affected until the thermometer adjusts to ambient temperature, which could take as long as 30 minutes. This is only a problem if you are taking absolute readings; comparative readings to surrounding objects are still possible.

An infrared camera uses technology found in the infrared thermometer and multiplies the display. In other words, an infrared thermometer has one pixel camera with a digital display. An infrared camera has hundred or

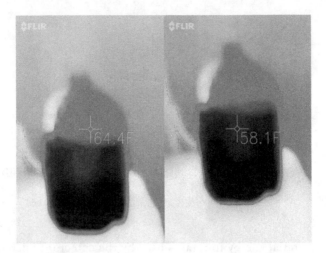

Figure 3.20 Image from an infrared camera showing relative temperatures of a one-gallon (4L) glass container. The liquid level can be determined as cooler than the headspace, both of which are cooler than the surrounding air. Note the lighter color of the handle, which is still warm from the hand that carried it. The digital temperature display is accurate within the aiming point while the colored display is relative to the aiming point. Color display is relative to the observed area with cooler temperatures represented by cool colors (black, purple, blue) through warmer temperatures by warm colors (orange, yellow, white).

thousands of pixels with a color display. Usually, the center pixel in a color display shows a digital temperature.

Temperatures appear in a relative visual display with blue/purple being colder and yellow/white being hotter as shown in Figure 3.20. An infrared camera can be used to determine the level of contents in a container as well as temperature of the container wall. The camera cannot "see into" a container; it can only determine the temperature of the surface of an object. Specifically, it cannot determine temperature behind a glass container or window.

Infrared cameras are becoming more affordable. Some are adaptable to a cell phone or unmanned aerial system (UAS).

Increased temperature in a substance could be an indication of a chemical reaction occurring in the container or the high reading could be caused by latent heat after moving the sample from a warmer area. Highly radioactive sources are hot and would have a higher temperature than their surroundings.

Conventional and electronic thermometers must make contact with the material. These are more useful in a laboratory where a smaller volume of unknown material close to the operator is not as dangerous as a large volume of material in an uncontrolled environment.

There are many models of electronic thermometers available. Choose one with simple calibration requirements, fast response, and a wide range of scale. A highly precise thermometer is not necessary unless it is being used for melting point, freezing point, or flashpoint determinations. If using a conventional glass thermometer, choose one that does not contain mercury; breaking the glass and spilling the mercury could be more dangerous than the material you are characterizing.

Mass Spectrometry

Mass spectrometry (MS) is a method of measuring the mass-to-charge ratio of ions. A sample material is ionized, which is then separated by mass. A quantitative determination is made by measuring the intensity of the ion flux. A mass spectrometry instrument basically consists of an ionization source, a mass analyzer, and a detector system.

MS instrumentation suitable for field use is designed as a stand-alone detector that samples air. Ionization of gaseous material can be achieved at atmospheric pressure. The ionization source ionizes the sample material by one of several techniques, depending on the application. Ionization may be achieved by electric or chemical sources, which are typically used for analysis of gases and vapors. Liquid and solid materials are often ionized by electrospray or matrix-assisted laser desorption/ionization (MALDI). Other ionization sources are in use based on application. Once ionized, the sample is moved by magnetic or electric field to the mass analyzer.

The mass analyzer separates ions by their mass-to-charge (m/z) ratio. The mass-to-charge ratio means two variables determine the behavior of the ion in the analyzer. Simplistically, ions are moving while being affected by the inertia of their mass as well as their charge near a magnetic or electrically charged surface. Analyzers can use static or dynamic fields of electricity or magnetism. Since more than one combination of analyzer is available, they are sometimes used in series to improve results.

The detector senses the induced charge or the current produced when an ion is nearby or in contact with the detector. When coupled with an amplifier and signal processor and computer, a visual mass spectrum is produced, which can be used to identify unknown mixtures of ions when matched with a library of known spectra.

There are several types of mass analyzers. Sometimes they are combined to improve resolution, compare spectra, etc. The main types of mass analyzers are presented below. Each type has advantages and disadvantages, and they are often customized to specific applications. Most applications are found in the laboratory; field portable instrumentation is not as prevalent, but development is continuing.

A sector mass analyzer uses an electric and/or magnetic field to influence the path of ions through the analyzer (Figure 3.21). The field will have a greater effect in moving lighter and highly charged ions and will separate

Sector Mass Spectrometry Function

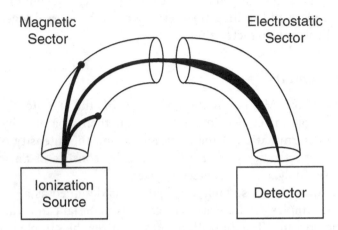

Figure 3.21 Ions enter the analyzer from the ionization source. In the magnetic sector ions are separated by mass; high mass ions are deflected less than low mass ions, and only the mass of interest is able to enter the next sector. Sorting ions in this manner is adjustable based on the strength of the magnetic field. The ions enter the electrostatic sector where ions of the same mass-to-charge ratio undergo energy distribution compensation and are focused on the detector.

ions based on their characteristics. Segregated ions will exit the analyzer focused into beams. Ions can be analyzed by the location or sector of the detector they strike. The electric or magnetic field can also be used to deflect the path of undesired ions away from the detector surface so only ions meeting certain specifications are examined. It can also be used over time to scan through a mix of ions.

A time-of-flight mass analyzer (MS-TOF) uses a uniform electric field to accelerate ions, which will move according to their charge and mass (Figure 3.22). If the field is negatively charged, positive ions will be accelerated toward the detector, and vice versa. The field works like an "electric gate," and time of flight is measured from the opening of the gate until the ions arrive at the detector. The lightest ions arrive more quickly and the heavier ions arrive more slowly. There may be pauses in flight time between groups of ions. The resulting spectrum of peaks and valleys depicts the arrival time and the relative abundance of ions.

A quadrupole mass analyzer moves ions through oscillating electrical fields within a radio frequency quadrupole field. The effect is selective stabilization or destabilization of certain ions. This method can selectively filter ions by mass.

The quadrupole ion trap mass analyzer is similar to the quadrupole mass analyzer, except the ions are held and selectively ejected as the radio frequency

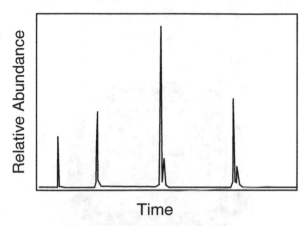

Figure 3.22 Hypothetical MS-TOF spectrum showing relative abundance of ion type and arrival times at detector.

potential is ramped. Other variations exist for the ion trap. A linear version of the quadrupole trap isolates ions in two dimensions instead of three.

Cyclotron mass analyzers induce ions into a pulsing cyclotron orbit within the analyzer to determine the mass-to-charge ratio of ions based on the ion cyclotron frequency within a fixed magnetic field. The ions are trapped within a magnetic field and charged electric plates. An oscillating electric field perpendicular to the magnetic field moves the ions into a larger cyclotron radius.

The ions begin to move in phase, producing a unique signal that can be processed with application of Fourier transformation (a method of separating several waves of different frequencies). The Fourier transform mass analyzer differs significantly from other mass spectrometry techniques in that the ions do not strike a detector but only pass nearby. The result is the ion masses are determined by frequency, not by focusing a beam on a certain location on a detector or by measuring time of flight.

An orbitrap mass analyzer is another type of cyclotron mass analyzer that uses an electrostatically charged spindle to trap ions, as opposed to the traditional Penning trap described with the Fourier transform mass analyzer. As the ions orbit the spindle, they oscillate along the spindle axis. Detector plates similar to those described with the Fourier transform mass analyzer collect information, which is processed by Fourier transformation to generate a spectrum.

High-Pressure Mass Spectrometry

Hand-held high-pressure mass spectrometers are available in the form of the M908 and MX908 from 908 Devices (Figure 3.23). These systems are

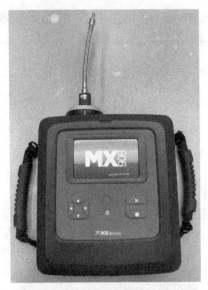

Figure 3.23 Example of a high-pressure mass spectrometer, MX908. The instrument can be configured to test airborne samples (top) at a rate of 3 liters per minute, or nonairborne samples using a thermal desorption accessory (bottom). The strip with the "MX" cut-out is the swab inserted in the thermal desorption unit. (Image: Kjellt Zomer, GATE Specialties, NL. With permission.)

quadrupole ion trap mass spectrometers that operate on the same principals as larger laboratory ion trap mass spectrometers. Field instruments contain microscale ion traps are able to operate at higher pressures than lab instruments, which improves portability by reducing the vacuum pumping system size and power.

A quadrupole ion trap MS system traps ions in a three-dimensional structure of electrodes. The ions naturally segregate by mass-to-charge ratio in the trap. Then the ions are ejected in order of mass to charge from smallest to largest, providing a readout of the abundance of ions at a given mass to charge. The M908 and MX908 systems identify ions from approximately 45 to 500 m/z.

The resulting mass spectrum can then be compared to the expected spectrum based on reference materials or operator knowledge.

Laboratory mass spectrometers often operate at 1×10^{-9} atmosphere (essentially space vacuum) and can take hours to start up. HPMS systems operate at about 0.001 atm. The small vacuum volume and higher pressure design make rapid start-up possible for field use.

Like all mass spectrometers, these systems have to first make ions from the sample, and these field instruments use a corona discharge ionization source. Vapors are simply drawn into the ion chamber. Involatile samples like explosives or drugs are first heated with a thermal desorber to liberate a minute amount of vapor into the ionization chamber. Fragments are collected in the quadrapole ion trap and then streamed onto the detector surface. Both positive and negative mass spectra are collected, which improves the selectivity of the resulting analysis (Figure 3.24).

The instrument detects and identifies chemicals continuously in real time when sampling vapor in air. When analyzing solids or liquids with the thermal desorber the system takes approximately 30 seconds to heat/analyze the sample.

A complex sample can produce ions from numerous species that may be simultaneously measured and overlap. MX908 uses a technique called collision induced dissociation (CID) to break the complexity of a mixture into component parts. The process fragments the main ions from the sample into smaller, lighter ions. This provides a multilayer analysis where separate mass spectra are obtained for positive and negative ions, as well as differing levels of fragmentation. The use of CID enables identification of targeted material in, for example, heavily cut street drugs or aged chemical warfare agents

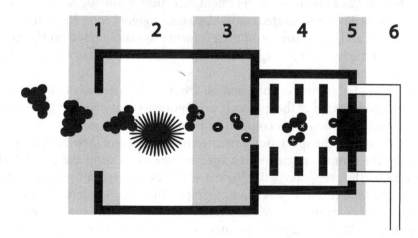

Figure 3.24 Schematic representation of HPMS. (1) Sample molecules are collected from the atmosphere or thermal desorption unit. Corona ionizer (2) fractures molecules into ions (3). Positive and negative ions are suspended in the quadrupole trap (4). Ions are counted on a collector surface (5). Vacuum is maintained to prevent interference from air (6).

while ignoring interference from common environmental materials or inert ingredients. Materials like hydrocarbons, sugars, starches, and oils are intentionally invisible to HPMS systems by configuration.

HPMS systems have a custom-built reference library with mass spectra corresponding to positive/negative modes and different CID sequences. As an ion trap mass spectrometer, the resulting spectra are directly comparable against some reference sources. Some online tools can produce a "predicted" mass spectrum for the MX908 from a chemical structure.

The MX908 uses some of these techniques in its embedded software to produce "class" alarms. For example, the instrument will alert the operator to fentanyl analogs that aren't exactly represented in the system library, rather than provide a "no match" result. Unfortunately, the NIST mass spectral library is not relevant or comparable for HPMS, as it only applies to mass spectrometers using electron impact ionization sources.

Gas Chromatography-Mass Spectrometry

Gas chromatography-mass spectrometry (GC-MS) instruments utilize, as the name suggests, a gas chromatograph in line with a mass spectrometer. The gas chromatograph separates molecules based on their volatility as they diffuse through a column of gas and information is obtained on the time it takes (retention time) for the individual components that make up the sample to migrate to the end of the gas column. As a separation technology, a gas chromatograph may be coupled to any number of detection instruments. It is included here as a common form of field detection technology.

As the sample is separated, individual components enter the instrument sequentially and are ionized and then information is obtained on the mass-to-charge ratio by the mass spectrometer. Each molecule has a unique fragmentation spectrum.

A computer compares information from both instruments to obtain a reliable identification of the molecular components that make up the sample. A GC-MS instrument will compare the relative concentrations among the atomic masses in the spectrum generated by the detector. The computer will compare the sample spectrum to a spectral library to see if the characteristics match. The computer will also correlate the spectrum to the retention time of the sample in the gas chromatograph. The integration of GC with MS allows a higher degree of identification than if the technologies were used separately. GC alone can produce unclear results if two or more components have similar retention times. MS alone can identify a pure substance, but mixtures can produce spectra that are not readily recognizable due to the influence of more than one substance. Additionally, some molecules produce ionization patterns that are similar to other molecules. Integration of the two technologies allows the sample to be physically separated and then analyzed by retention time and mass spectrometry. This correlation leads to a higher degree of confidence.

Other GS-MS strategies that are more specific and analytical may be used in a laboratory, but for field use, you're depending on the computer's ability to recognize inputs and match them to a spectrum already present in the machine's library. Some libraries may be customized to a substance of interest based on controlled observation and recording of a known sample concentration.

The gas chromatograph may require one or more carrier gases for the gas column. These gases are specified by the manufacturer and have specified purities. The gas may be pure scientific grade nitrogen, helium, hydrogen, pure air, or others. Humidity and contaminants must be controlled. Cycle time for a sample may be about 90 seconds. During that time the GC column is flushed with carrier gas and then the sample is injected into the column. After the sample has cleared the column, it is flushed again. A detector may employ more than one column, and cycle times can extend to an hour or more for slow moving compounds. Back flushing clears the column of slow moving compounds after faster moving compounds of interest have been detected and allows the GC to analyze a new sample.

The GC column is heated according to the temperature ramp programmed as part of the method the instrument is running. The steadily increasing temperature of the column allows chemicals to volatize (elute) out of the column at their retention time correlated to volatization temperatures. Many chemicals have similar boiling points resulting in coeluted peaks along a chromatogram. Technologies such as mass spectrometry allow for specific identification of chemical molecules even when coeluted on a chromatogram. A heated column makes it less likely that heavier sample components will condense on the column rather than travel through it.

GS-MS detectors require a long startup time. If the detector is to be ready for immediate use, it must be left on. Even in storage a slight flow of carrier gas must be maintained so the column does not become contaminated.

A gas chromatograph can be combined with detection technology other than a mass spectrometer, such as a photoionization detector. Two examples are the Photovac Voyager Portable GC and the Inficon HAPSITE® Field Portable System from INFICON.

Ion Mobility Spectrometry

Ion mobility spectrometry (IMS) sorts ions by their mobility. Mobility is determined by timing ion response in an electric field and a known atmosphere within a detector tube. An ion will be attracted to an appropriately charged surface through the electric field at a calculable velocity and will arrive predictably. Resistance to movement occurs when the ion collides with gas molecules; large ions will have more collisions and have reduced mobility. Several gases may be used within the tube, including atmospheric air. IMS instruments operate with a variety of gases in the tube and within several temperature and pressure parameters; portable field use requires the selected instrument operate in air within the range of ambient temperature and pressure expected for typical use.

The IMS monitor draws a sample into a detection chamber that may be protected by a thin membrane, which is impermeable to water. Persistent material (some chemical warfare agents and other solid and liquid hazardous material) may need to be heated in order to vaporize the substance and make it mobile in air. The membrane serves to keep dust and other particulate matter out of the detection chamber as well as reduce the amount of water vapor that enters with the sample. The membrane allows sample material to permeate through it before detection occurs. This means an increase in detection time and the nondetection of compounds that are impermeable to the membrane.

When a sample enters the detector, it is ionized by an energy source. The source may be UV light, electron spray, radioactive, or any other source capable of ionizing a wide range of compounds. A radioactive source is the most efficient source for field portable equipment and is common in military use. This radioactive source requires compliance with strict regulatory requirements. Civilian use IMS monitors often use a nonradioactive method of ionization, which requires more robust battery performance.

The ionization source fractures sample molecules into positive and negative ions at an electronic gate. The gate is pulsed to allow passage of positive ions for a determined amount of time and then the field on the gate is reversed to allow passage of negative ions for a determined amount of time (Figure 3.25).

The ions "drift" very quickly to collector surfaces that attract the ions by opposite charge and are located beyond the gate. The time of travel is measured from the moment the gate opens until the ions are detected on a collector plate in microseconds to milliseconds. As the ions travel through the detection chamber, they exhibit three characteristics: drift, diffusion, and collision.

Drift is the travel of the ion from the gate to the detection surface. Drift is accelerated by higher and opposite charge in the detection chamber. If the voltage applied to the sample is different than that used to generate the library signature, the spectra will differ.

Diffusion is the dispersion of ions from higher to lower concentrations, much like smoke diffusing as it moves downwind from a fire. Diffusion is not consistent and is influenced by properties of the sample compound and temperature of the sample. Diffusion causes broader ion peaks in the spectrum as drift time increases.

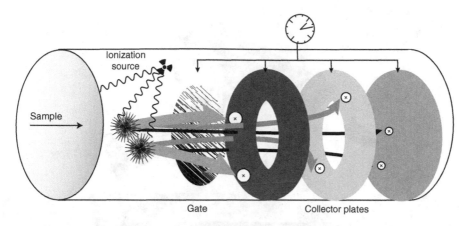

Figure 3.25 Filtered air enters the ionization chamber, where sample material is ionized. An electric gate admits either positively or negatively charged ions. Charged collector plates influence the drift of the ions and their arrival time from the gate is recorded. A spectrum is developed from the signal received from the collector plates relative to time of travel.

Collision occurs when ions strike air or other gas molecules within the detection chamber. Collision will increase drift time and alter the sample signature. Collision rates can be affected by air density, which is influenced by atmospheric pressure, temperature, and humidity. These variables might be accounted for through the use of sensors and software in instruments as the technology matures.

The detection chamber contains areas that attract the ions. When ions meet the ion collector, the ion is neutralized and an electric signal is generated. The signal is sent through a processor and compared to representative spectra in the library. Variables in drift, diffusion, and collision combine to influence the spectrum by changing the amplitude and width of a wave as well as shifting it in time. If a match occurs, an alarm sounds or a compound name is specified.

The representative spectrum of a compound held in the library may not match the sample if variables in drift, diffusion, and collision are different. Software or a skilled operator may be able to interpret the differences between the sample spectrum and the stored spectrum.

IMS instruments measure ion mobility in tens of milliseconds. When this feature is considered with the ease of use, relatively high sensitivity, and highly compact design, IMS is considered a useful tool for the field detection of explosives, drugs, and chemical weapons, and characterization of unknown material.

Some monitors may use simpler algorithms that alert the user to the presence of a sample that contains certain characteristics. For example, an IMS monitor (Figure 3.26) could be programmed to use audible and visual alarms

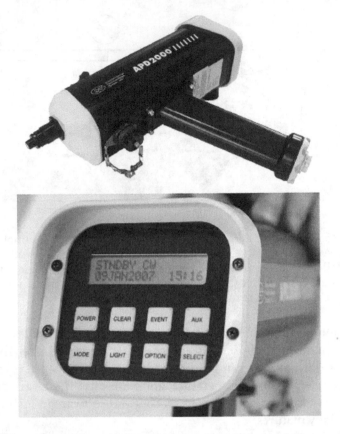

Figure 3.26 Advanced Portable Detector (ADP) 2000 ion mobility spectrometry air monitor from Smiths Detection.

to alert the operator to the presence of phosphorus atoms, which are contained in G- and V-nerve agents, and sulfur atoms, which are contained in HD (mustard) and V-nerve agents. The assumption is that compounds containing phosphorous and sulfur are not common in the air, and if the monitor is being used in a possible chemical warfare agent setting as designed, the sample should be assumed to be a chemical warfare agent.

An IMS monitor will not produce a clean spectrum for a known compound if the concentration is too high. If the sample concentration is too high, the ionization source will not be able to ionize the entire sample and whole or partially ionized fragments of molecules will enter the detection chamber and influence the signal being generated by the ion collector.

If more than one substance is present in the ionization chamber, the more easily ionizable substances will be highly misrepresented while other portions may go undetected. The spectrum will be a composite of overlapping and altered spectra of the sampled compounds. Software might be able to separate and identify some mixtures.

Examples of field portable IMS detectors are HazMatID, Sabre 4000, CAM-2, RAID-M 100®, GID-3®, and LCD.

Sandia National Laboratories Micro Analytical Systems is developing a miniaturized IMS detector using an inexpensive drift tube made from low temperature cofired ceramics (LTCC), which provides more chemically inert internal surface along with integral potential resistors. This design eliminates up to 150 individual parts from current designs. The LTCC miniaturized IMS detector is producing results similar to conventional IMS instruments, but as of this writing has not become commercially available. It is expected to have application in detection of chemical agents, explosives, biological material, drugs, and other hazardous materials.

An IMS monitor can be a very useful tool in the field. However, if the sample concentration is too high or mixtures are involved, a skilled operator is required to prevent misdirection based on a produced library match.

Since the first edition of this book, many first responders have made know their opinion that hand-held IMS units were not meeting their needs in the field. However, IMS does deserve a place at the table. Idaho National Laboratory (INL) published results in 2008 of a survey of 174 first responders who owned a hand-held IMS. The survey identified the following observations:

- The most common IMS unit used by respondents was the Advanced Portable Detector (APD-2000), followed by ChemRae, Sabre 4000, Sabre 2000, Draeger Multi IMS, Chemical Agent Monitor-2, Chemical Agent Monitor, Vapor Tracer, and Vapor Tracer-2.
- The primary owners were hazmat teams (20%), fire services (14%), local police (12%), and sheriffs' departments (9%).
- IMS units are seldom used as part of an integrated system for detecting and identifying chemicals but instead are used independently.
- Respondents are generally confused about the capabilities of their IMS unit. This is probably a result of lack of training.
- Respondents who had no training or fewer than 8 hours were not satisfied with the overall operation of the hand-held IMS unit.
- IMS units were used for detecting a range of analytes. The most common use was for detection of hazardous chemicals, followed by detection of explosives, illicit drugs, chemical warfare and nerve agents.
- Respondents who did not own an IMS listed prohibitive cost of equipment as the main factor for not having one.
- Respondents who were highly satisfied with the overall operation of the hand-held IMS obtained the IMS through a direct purchase. In comparison, the respondents who were not satisfied had obtained the hand-held IMS through a DHS grant.

High-Field Asymmetric Waveform Ion Mobility Spectrometry

High-field asymmetric waveform ion mobility spectrometry (FAIMS) is a newer technology and a variant of conventional Ion Mobility Spectrometry (IMS). It is used to separate gas phase ions at standard temperature and pressure as well as slightly higher and lower temperature and pressure.

A mixture of ions of varying size and type is introduced between two metal plates that have been charged to voltages up to 15,000 volts/cm. As in IMS, the characteristics of some ions will cause them to drift into the metal plates and neutralize. The characteristics of certain ions will cause them to remain "suspended" between the plates or to drift more slowly toward a plate. Altering the charge on the plates will cause the ions to move differently.

FAIMS alters the charge on the plates by using an asymmetric waveform, which causes the ions to separate based on the characteristics of the asymmetric waveform. Asymmetric waveform means the positive voltage may be higher and longer lasting than the weaker and shorter lasting negative voltage, or vice versa. The alternating and varying charge causes the ion to move in a "stair step" route rather than moving continuously toward a plate (Figure 3.27).

Some ions will experience a mobility that increases with electric field and other ions will experience a mobility that decreases with electric field. These ions will migrate in opposite directions between the plates. Additionally, ions will have varying degrees of mobility – that is, they will move slowly or quickly compared to other ions. If a small DC voltage is applied to either of the plates, the ion mobility will be affected by the magnitude and polarity of the DC charge. This means the ability of the instrument to detect a particular ion is dependent on varying asymmetric waveform adjustments and the compensating DC voltage.

A mixture of ions introduced between the plates immediately reacts to the asymmetric waveform by separating and moving at various rates toward the plates as the gas carries the ions toward the detector. Adjustment can be made to the separating mix by varying the DC compensating charge.

The terminal velocity of the ions traveling through the gas is roughly proportional to the strength of the electric field. This proportionality changes at high electric field, is compound specific, and can be used to segregate ions in two dimensions. When coupled with an ion trap, the ions can be segregated in three dimensions and a corresponding three-dimensional spectrum can be produced. A three-dimensional spectrum can produce a more detailed result and help you determine components of a mixture with improved accuracy. FAIMS can provide a more detailed chemical fingerprint, resulting in a higher degree of confidence in the chemical identification and a lower false-positive rate.

If the flat plates are replaced by cylinders, the ions can be focused to a more central point. The flat plate system distributes ions around a location

Oscillation of Ions in FAIMS

Direction of ion drift and detector

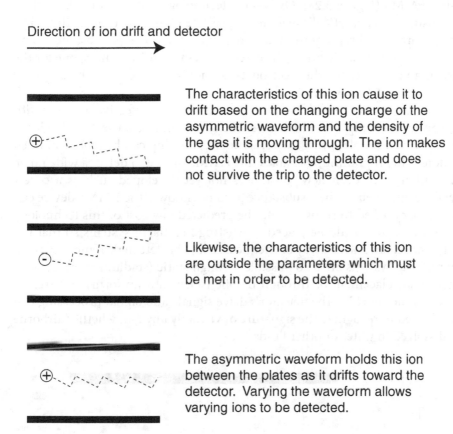

The characteristics of this ion cause it to drift based on the changing charge of the asymmetric waveform and the density of the gas it is moving through. The ion makes contact with the charged plate and does not survive the trip to the detector.

LIkewise, the characteristics of this ion are outside the parameters which must be met in order to be detected.

The asymmetric waveform holds this ion between the plates as it drifts toward the detector. Varying the waveform allows varying ions to be detected.

Figure 3.27 Due to charge and mass characteristics of each ion, the ion will drift toward one plate or the other or remain "neutral" between the plates. Ion drift can be affected by altering the asymmetric waveform.

due to influence by diffusion, ion to ion repulsion, gas flow, and gas turbulence. When the flat plate geometry is replaced by a radial system of two cylindrical plates, one inside the other, the ions will exit the plates, tending to be focused at a certain radius. This results in a more concentrated flow of ions and helps produce detection at lower concentration. This radial system, a "tube in a tube" design, exists in laboratory instrumentation but is not known to be field portable.

Owlstone Nanotech has used micro- and nano-fabrication techniques to create a proprietary chemical detection system purportedly 100 times smaller and 1000 times less expensive than existing technology. This sensing technology was designed to meet the specifications from the U.S. Department of Defense.

This sensor is a silicon chip that provides a chemical sensing mechanism using FAIMS (Figure 3.28). Owlstone's technology enables miniaturization of sensors with analytical capability at projected lower cost, the ability to be programmed and reprogrammed to detect a wide range of substances, and high selectivity and sensitivity. The device uses two silicon chips, one for the sensor and another for the electronics and ionization source. The chips are thin and about 1 cm square.

This technology enables a miniaturized and cost-effective detector with low power consumption, which is beneficial for hand-held use outside a laboratory. When coupled with a wireless device, the chip can be reprogrammed remotely. Reprogramming enables this FAIMS device to detect a wide range of substances or even new substances not yet developed such as a covert nerve agent; when the new substance becomes known, the FAIMS device can be reprogrammed to recognize it. The predicted low cost of this technology suggests sensors could be placed in existing fire alarm systems so that not only smoke, but a wide array of chemicals could be detected. Temporary area monitoring systems are also possible through static (stadium, parade, demonstration) placement or through mobile (sewn into uniforms, adhered to vehicles) monitoring. The detector's drive signals and signal processing can be adjusted to recognize the signature of virtually any gas, whether airborne or dissolved in water or other fluids.

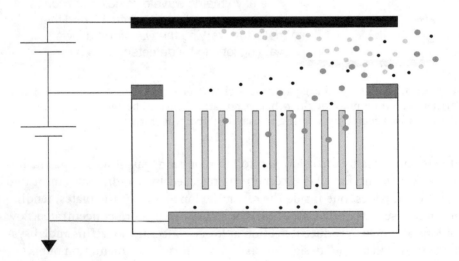

Figure 3.28 In this schematic representation of a miniaturized FAIMS detector, ions enter the sensor from the right and are driven by electric charge through a set of charged plates. The sensor acts as a reprogrammable filter, which separates and identifies chemicals according to their characteristic mobility. The sensor "filters" out background chemicals that do not have the mobility specified by the program. Ions of interest then move to the detector plate, which generates a signal that is processed into a spectrum.

Owlstone has developed a field portable instrument based on FAIMS, called NGCD (next-generation chemical detector).

Photoionization

Photoionization detectors (PID) are nonselective monitors used mainly to quantify a known volatile organic compound (VOC). They may also be used to screen a sample for VOCs and some inorganic compounds.

A PID uses a light source, typically a UV lamp, to ionize molecules. A molecule with a loosely held electron may be struck with a photon of a particular energy level. The photon drives the electron away and leaves the molecule as a positively charged ion. If an electron is held more tightly, the photon may not have sufficient energy to drive the electron away and the sample would not ionize or be detected. The affinity of the electron for the molecule is measured in potential with units of electron volts (eV).

PID UV lamps typically produce UV light in certain wavelengths. Shorter wavelength radiation contains higher energy. The shorter the wavelength of the light source, the more powerful the ionization potential. The ionization potential of lamps typically ranges from 8.3 to 11.6 eV (Table 3.14). Most PIDs utilize a 10.6 eV lamp, which represents a compromise between

Table 3.14 Compounds Detectable by PID

Compound	IP (eV)	9.5 eV	10.6 eV	11.7 eV
			Lamp	
Aniline	7.70	Y	Y	Y
Naphthalene	8.12	Y	Y	Y
Xylene	8.56	Y	Y	Y
Methyl bromide	10.54	N	Y	Y
Nitrogen dioxide	9.75	N	Y	Y
Phosphorous trichoride	9.91	N	Y	Y
Propane	11.07	N	N	Y
Tetraethyl lead	11.10	N	N	Y
Phosgene	11.55	N	N	Y
Hydrogen cyanide	13.60	N	N	N
Nitric acid	11.95	N	N	N
Sulfur dioxide	12.30	N	N	N
Major Components of Air				
Nitrogen	15.58	N	N	N
Oxygen	12.08	N	N	N
Carbon dioxide	13.79	N	N	N
Water	12.59	N	N	N

Note: Y = Detectable; N = Not detectable.

effectiveness, cost, and lifespan of the lamp. Lamps may be changed in the field, but a period of stabilization and calibration is necessary that would significantly slow results (Figure 3.29).

A negatively charged detector surface attracts the ion and a signal is generated when the two make contact. A signal processor passes the information to a display that reads the total amount of ionized material detected. The PID is nondestructive – that is, the sample is converted from an ion back to its original state when the charge is neutralized on the detector surface.

Significantly, a PID will produce a reading that is the sum of all compounds that are ionized by the UV lamp in the detection chamber. Since many ionizable compounds may be present in a sample, a PID will not necessarily produce a quantitative measurement of a particular compound. For example, the PID may indicate a reading of 100 ppm, but it is up to the operator to determine if the sample contains 100 ppm of a pure compound or some combination of ionizable compounds in unknown proportions.

If a sample contains a single, known compound, the PID may be calibrated with a gas standard and used to determine the concentration of the sample gas based on a response factor. The response factor is dependent on both the sample and the calibration gas.

Examples of photoionization detectors are HNu 102®, ppbRAE® Plus VOC Detector Monitor, and Passport® PID II Organic Vapor Monitor. A PID

Figure 3.29 The Passport PID Organic Vapor Monitor photoionization detector from MSA. This unit is shown with the sampling probe attached. The photoionization lamp and cap are pictured on the right.

may be useful in screening for total VOCs and a few other compounds, but is not generally useful in identification of unknown compounds.

Flame Ionization

A flame ionization detector (FID) is similar to a PID, but it uses a hydrogen flame instead of a UV lamp as an ionization source. FIDs are nonselective monitors used mainly to quantify a known volatile organic compound (VOC). They may also be used to screen a sample for VOCs.

The hydrogen flame in an FID burns between 2000 and 2500°C and produces ions from organic samples. The ions are produced proportionally to the concentration of all organic compounds in the sample.

A negatively charged detector surface attracts the ion and a signal is generated when the two make contact. A signal processor passes the information to a display that reads the total amount of ionized material detected. Unlike the PID, the FID is destructive – that is, the sample is destroyed to create ions, which are then neutralized on the detector surface.

An FID will produce a reading that is the sum of all compounds that are ionized by the hydrogen flame in the detection chamber. Since many ionizable compounds may be present in a sample, an FID will not necessarily produce a quantitative measurement of a particular compound. For example, the FID may indicate a reading of 100 ppm, but it is up to the operator to determine if the sample contains 100 ppm of a pure compound or some combination of ionizable compounds in unknown proportions.

If a sample contains a single, known compound, the FID may be calibrated with a gas standard and used to determine the concentration of the sample gas based on a response factor. The response factor is dependent on both the sample and the calibration gas. An FID may be useful in screening for total VOCs and a few other compounds, but not in identifying them. Selectivity is possible when the FID is coupled with a separation technology, such as a gas chromatograph.

Examples of FIDs are Photovac MicroFID® and Signal Group 3030PM®. Some monitors combine PID and FID capability into a single unit, such as the Foxboro TVA-1000B Toxic Vapor Analyzer® from Thermo Environmental Products.

In a laboratory setting, an FID or PID may be coupled with a gas chromatographic interface to separate components of a sample and increase selectivity, but this technology is not currently practical in the field for immediate characterization of unknown compounds.

Surface Acoustic Wave

Surface acoustic wave (SAW) technology may be used to identify unknown substances.

A piezoelectric crystal will produce a measurable electric charge when it is subjected to pressure. This technology is used in the microphone of a telephone to convert the pressure generated by the sound waves of your voice into an electric signal, which is in turn transmitted to the receiver's speaker. A piezoelectric crystal will also slightly change shape if an electric current is applied to it.

A SAW monitor (Figure 3.30) uses a flat piece of piezoelectric material that is subjected to alternating current. The rapid cycling of the current causes waves to pulse across the surface of the piezoelectric material. This wave pattern is the baseline pattern against which changes will be measured. When a molecule is placed on the surface, the wave pattern is altered slightly, similar to the slightly changed wave pattern around a floating ocean buoy. A second piezoelectric substrate is used to detect the changing wave pattern and converts the wave pattern into an electric signal, which can be processed and read by a computer.

Since the piezoelectric material is a hard and smooth crystalline surface, a thin coating of material is applied over the surface to collect vapor. The coating is a polymer with available monomer bonding sites extending above the surface. As the sample passes over the coating, three results are possible:

- The sample molecule may bond to the monomer and affect the SAW pattern.
- The sample molecule may be absorbed into the polymer and affect the SAW pattern.
- The sample molecule may not be soluble in the polymer and may not be able to be bonded to the monomer, in which case the substance is not detected.

Figure 3.30 HAZMATCAD surface acoustic wave detector by MSA. HAZMATCAD incorporates electrochemical cells to expand its capability.

Environmental factors can affect the baseline wave pattern as well as the polymer coating composition. Some SAW monitors utilize an array of sensors. One is used to subtract current environmental effects. A combination of SAW sensors with differing polymer coatings can be used to produce a set of results for the same sample. The results are compared to a library of results by a computer and an alarm is produced based on a match. Similar to other library comparisons, mixtures can confound the effort of identification because results must be added or subtracted from the baseline wave pattern.

After testing, the sample must be cleared from the polymer surface. Usually a heat source is applied to vaporize the sample from the polymer and it is flushed from the area by clean air. Subjecting a polymer to a sample that cannot be cleared may disable the monitor. A lifespan of about a year is typical of the polymer coatings.

The SAW MiniCAD mk II from Microsensor Systems uses surface acoustic wave technology, as does the ChemSentry™ 150C Detector System from BAE Systems.

Flame Photometry

Flame photometry is a method of burning material and measuring light emitted by hot atoms as they cool (Figure 3.31). Atoms and molecular fragments of thermally degraded substances emit unique light signatures. These signatures may be seen approximately by the human eye as colors and the concentration of the sample in a flame is seen as brightness.

A flame photometry detector (FPD) is capable of "seeing" more colors than the human eye, such as the "invisible" flame of hydrogen or some alcohols. An FPD is also capable of filtering undesirable light and detecting specific wavelengths far beyond the ability of the human eye. The capability to detect a specific frequency over a wide range of low-intensity light is the principle behind the flame photometer. While an FPD will not identify a specific compound, it will identify some of the "building blocks" that emit characteristic light. Knowing which elementary components are present can help the operator deduce characteristics of the unknown material (Table 3.15).

An FPD draws the sample into a hydrogen flame and complex molecules degrade into simpler portions. The fragments may recombine in any number of combinations.

When an atom is heated and sufficient energy is added, an electron will move into a higher energy state. When the heat is removed and the atom cools, the electron will fall back to its base state and emit a photon with certain light energy characteristics. Specific light energy spectra are associated with specific molecular species, and the detector employs light filters to block extraneous light energy. The light emissions are detected by a photomultiplier tube, a device that amplifies the light signal, and passes it to a signal processor.

Figure 3.31 Colors are characteristically produced by hot or burning materials containing certain elements.

Table 3.15 Observed Coloration of Flame Produced by Some Elements

Element	Flame Color
Arsenic	Blue
Barium	Yellow-green
Boron	Green
Calcium	Orange-red
Copper	Green
Lead	Gray-blue
Lithium	Red
Potassium	Lavender
Sodium	Yellow-orange

The wavelength of the photon is measured by the FPD and the operator is notified by a readout or alarm. If too much heat energy is added by the flame, an electron is driven away and the atom ionizes. The heat source must be regulated based on the atom to be detected. For example, the AP2C FPD uses a hydrogen flame with a temperature of 2000°C–2100°C while monitoring for sulfur and/or phosphorous in chemical warfare agents and toxic industrial chemicals (Figure 3.32). It is important to maintain the flame temperature through consistent hydrogen air mixing so that sensitivity remains constant. A flame photometer may not be intrinsically safe for use in a flammable environment (the AP2Ce model is designed for use in explosive atmospheres).

Specifically, sulfur-containing molecules that can produce excited S_2 molecules in the heating process can be detected by their characteristic emission of 394 nm light waves. Phosphorus-containing molecules, which can produce excited HPO molecules in the heating process, can be detected by their characteristic emission of 526 nm light waves. Nonsulfur or nonphosphorus compounds might be able to emit the target wavelength of light and would be indicated as sulfur or phosphorous. Future FPD design may include filter systems that allow detection of other characteristic emissions, which would allow identification of other species.

The AP2C will produce an alarm for presumed G-nerve agent (GA, GB, GD, and GF) if phosphorous is detected in the sample. Likewise, an alarm will sound for presumed mustard gas (HD) if sulfur is detected. If both phosphorus and sulfur are detected, the monitor will produce an alarm for V-nerve agent, which contains both atoms.

This level of selectivity is useful when the FPD is used in a setting in which a known CW agent has been released or for initial screening of an area for suspected CW agent. It is also useful for the detection of other sulfur- and phosphorous-containing compounds such as phosphine and sulfur dioxide;

Figure 3.32 AP2C flame photometry air monitor by Proengin SA. The instrument is alarming for G-nerve agent after exposure to a training surrogate.

however, this also means the detector will alarm for very low levels of normally present material, such as perfume.

The manufacturer of the AP2C makes a similar FPD called the TIMs (Toxic Industrial Material) Detector. The TIMs Detector has higher limits of detection to avoid false alarms from normally occurring background values.

Both the AP2C and the TIMs detectors use a heater device to vaporize solid and liquid samples from low volatility materials.

The MINI-CAM is an FPD for benchtop use in the laboratory. It is often connected to a gas chromatograph column to increase specificity. The MINI-CAM is not practical for first responder field use; however, it has applications in laboratory triage methodology.

Infrared Spectroscopy

Infrared radiation (IR) is energy that has longer wavelength, lower frequency, and less energy than visible light. Infrared radiation increases with temperature. Night vision goggles work by converting heat energy (infrared radiation) into visible light projected from a viewer. A firefighter's thermographic camera functions similarly by depicting infrared intensity. These devices detect the emission of infrared radiation from an object.

All molecules vibrate, and each bond within a molecule vibrates at a slightly different frequency depending on the type of chemical bond. The pattern of all vibrational frequencies of a particular molecule is unique, a molecular fingerprint. An instrument able to discern a unique pattern can recognize one molecule from thousands of others. There are several technologies available to measure and characterize aspects of the vibrational spectrum of a material. Although several forms of vibrational spectroscopy exist, two main instrumental techniques are currently used to characterize the vibrational fingerprint of a sample: infrared absorption and Raman spectroscopy. These techniques are most useful in the detection of organic and organometallic molecules. Infrared absorption and Raman systems can use different instrumental configurations, the most common of which are Fourier transform spectrometers and dispersive spectrometers. The Fourier transform devices are much more prevalent in infrared absorption, while dispersive systems are more common for Raman scattering.

Infrared spectroscopy detects the degree of absorption of infrared radiation when a wide frequency IR beam is returned from the solid, liquid, or vapor sample. The characteristics of the source IR beam are known. When the beam strikes a substance, the substance may absorb certain frequencies of IR. The remainder of the beam strikes the sensor and a computer develops a spectrum. The spectrum is the original beam spectrum minus the frequencies absorbed by the sample. Two types of sensors are used to detect the sample spectrum. A transducer capable of detecting IR is the preferred field sensor. The other type of sensor detects sound generated when IR affects molecular vibration; however, background noise can distort the result and this sensor is not recommended for field use. The fact that both sensors can be affected by noise requires operation in a quiet area.

Symmetrical diatomic molecules, such as nitrogen and oxygen, do not absorb IR. Most other molecules will absorb infrared radiation at one or more wavelengths. This includes water and carbon dioxide, two common atmospheric constituents.

If a wide spectrum beam of IR is used to interrogate a sample, characteristic absorption patterns may be obtained and compared to a library of known spectra. Comparison of pure sample spectra to stored spectra produces a very specific result. If a known sample is mixed in a known solvent, a computer may use a simple subtraction function to identify both components of the mixture. Multiple component mixtures in varying concentrations become a challenge. Perhaps a skilled operator who can manipulate the sample and combine IR interrogations could gather some information about the unknown sample. The introduction of contaminants to a sample reduces the effectiveness of IR spectroscopy in field applications. In order to reduce response time for field use, a library may be limited to certain compounds of interest. A larger library will increase response time for an extended search but will greatly increase specificity. An example of an infrared spectrometer that is a nondispersive IR absorption instrument is the MIRAN SapphIRe Portable Ambient Air Analyzer (Thermo Environmental Instruments, Inc.).

Photoacoustic Infrared Spectroscopy

Photoacoustic IR spectroscopy is another infrared technology based on sound produced when a sample is irradiated with an IR beam. The IR beam is pulsed into the sample where characteristic wavelengths are absorbed. When the material absorbs the infrared energy, its temperature increases, and the resulting gaseous expansion creates a sound wave that is then detected by a microphone. There is no optical detector required to quantify the amount of infrared energy absorbed by the sample. Photoacoustic infrared spectrometers can be sensitive to external sound, vibration, and humidity and must be calibrated to each operating environment.

Fourier Transform Infrared Spectrometry

Fourier transform infrared spectrometry is more commonly referred to as FTIR. This technique uses infrared radiation, similar to the method described above, but the beam is split and an interferogram is produced. A complex mathematical process called Fourier transform (FT) is applied to the interferogram and a final IR spectrum is constructed for the sample being tested.

This technique, for a variety of reasons, significantly improves performance while reducing the time required for testing.

Most FTIR instruments use a Michelson interferometer invented in 1891. The Michelson interferometer schematic in Figure 3.33 illustrates how half of an IR beam is distorted while the other half remains as a reference. As the moving mirror changes position, the power output of the beam changes as the two halves are not synchronized. The beam passes through the sample and produces an interferogram. Modern instruments do not require use of the reference half and are single beam instruments.

A Fourier transform mathematical function is applied to the interferogram to produce a spectrum characteristic of the sample. Sample and background spectra are proportioned to eliminate the instrument response. This also helps eliminate water and carbon dioxide from the spectrum, and spectral subtraction can be used to further eliminate these two substances. The ability to rapidly scan a sample with varying forms of infrared radiation coupled with the ability to rapidly process data makes FTIR overall a more useful technology than conventional infrared spectroscopy.

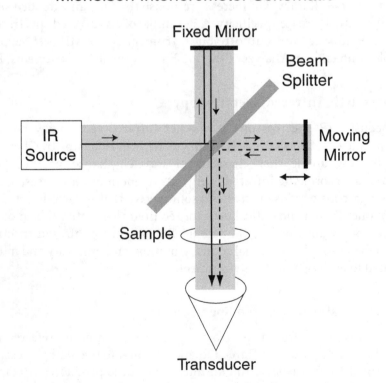

Figure 3.33 Schematic representation of the Michelson interferometer.

A high level of confidence can be assumed when FTIR is used to identify pure material from existing library spectra. Identification becomes much more difficult in mixtures containing low concentrations of the suspect material. An early evaluation of a portable FTIR analyzer recommended a simple solvent extraction technique for low levels of chemical warfare agents. The report states "complex spectrum from the mixture often leads to erroneous search results. Because unknown CWA found during an incident could potentially be mixed with dust or a liquid base, positive identification of such a material is important."

The report also states,

> If the solvent containing the agent is volatile and its evaporation leaves behind a concentrated sample of CWA, the instrument can readily identify the agent. This finding led to the chloroform extraction technique tried on VX mixtures. The findings were encouraging. Perhaps, by implementation of a simplified extraction process, the instrument would be much more useful for the identification of potential CWAs in a mixture.

Rapid advances in instrument capability have improved performance significantly. Simply placing a sample in a device and expecting it to instantly correctly identify the material is the goal, but is not necessarily a real-world occurrence. Physical and chemical properties encountered while characterizing or identifying an unknown material are realities of our world. Therefore it is important that anyone operating an instrument consider each situation individually and consider information other than the result the instrument produces. The instrument needs your skill and knowledge to work properly.

Examples of FTIR instruments are RAM 2000™ (Kassay), an open path FTIR system used to continuously monitor air for select contaminants; HazMatID Elite (Smiths Detection), a hand-held FTIR for solids and liquids; and Gemini® (ThermoFisher Scientific, Figure 3.34), a hand-held FTIR and Raman instrument for solids and liquids. Field use demands each instrument must be rugged and able to withstand decontamination. Many units now meet specifications of the U.S. military as described in MIL-STD-810G, *Environmental Engineering Considerations and Laboratory Tests*.

Raman Spectroscopy

C. Raman and K. Krishnan first observed the inelastic scattering of light in 1928. They projected sunlight from a telescope through a liquid or vapor and then used a lens to collect the sunlight scattered through the sample. A set of optical filters was used to observe the scattered light with frequency altered from the sunlight. This technique is commonly referred to as Raman spectroscopy.

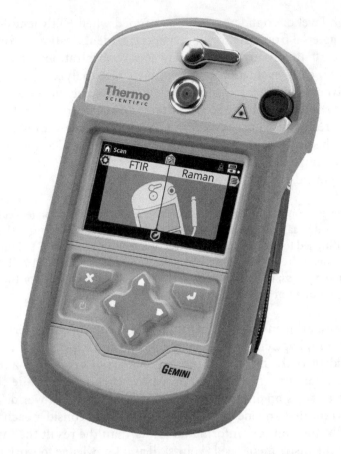

Figure 3.34 Gemini Raman and FTIR hand-held instrument by ThermoFisher Scientific. Raman and FTIR systems are packaged into a single platform and share a power source, computing system, touch keys, display and several other components. (Image: ThermoFisher Scientific, Tewksbury, MA. With permission.)

Three things can happen when a sample is subjected to a beam of infrared radiation:

- The beam can pass unaffected through the sample
- The beam can be partially absorbed (infrared spectroscopy)
- The beam can scatter
  - Most light scatters elastically with no change in frequency (Rayleigh)
  - About one of every 1,000,000 photons scatters inelastically and the frequency decreases (Raman)

Raman spectroscopy and the complimentary infrared absorption technology are the two main spectroscopy techniques used to detect vibrations in molecules. These techniques are used to fingerprint compounds as solids, liquids, and vapors from samples that can exist in bulk, as surface layers, or particles. IR spectroscopy uses a beam that contains a range of frequency and "missing" sections of range are detected. Raman spectroscopy uses a beam that contains only a single frequency and characteristic scattering caused by molecular bonds is detected for that particular frequency. In short, about one of every ten million IR photons directed into the sample will distort and polarize an electron cloud in a molecular bond. This distorted state is unstable and a photon is quickly ejected at a different frequency (Figure 3.35). With intense radiation and controlled conditions, the photons with shifted frequency, known as Raman shift, can be detected. Each type of molecular bond produces a characteristic Raman shift and every chemical compound produces a pattern of Raman shifts dependent on the type and location of its molecular bonds. A spectrum of these shifts is compared to a library of known spectra and can be used to identify the compound.

The most intense Raman scattering is observed from electron clouds that can be vibrationally distorted in a symmetrical manner. This condition makes Raman spectroscopy sensitive to detection of compounds with induced symmetrical vibration. In contrast, infrared absorption is most intense when asymmetrical vibration is present. When the two technologies are used in a complementary manner, detection is enhanced.

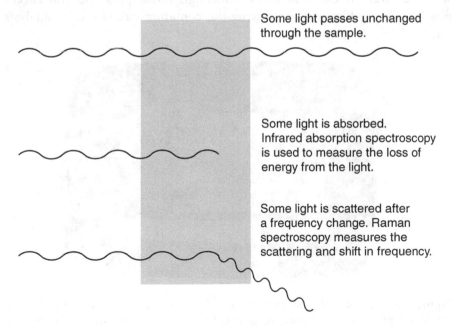

Some light passes unchanged through the sample.

Some light is absorbed. Infrared absorption spectroscopy is used to measure the loss of energy from the light.

Some light is scattered after a frequency change. Raman spectroscopy measures the scattering and shift in frequency.

Figure 3.35 Raman spectroscopy measures a change in frequency. The difference in frequency corresponds to an infrared frequency; however, the change may be in a near-infrared, visible, or ultraviolet frequency.

Raman spectroscopy can also develop spectral peaks for groups of atoms within a molecule. The ability to detect groups of atoms depends on the degree of symmetric distortion on bonds within the group as well as the geometry of the molecule and the group location relative to other atoms and groups. Because of these variables, groups can be detected in a band of frequency in which a peak can occur. Extenuating conditions can produce a peak outside the band. Other samples can produce very specific peaks. At least one field portable Raman spectrometer displays results based on probability after detecting signal within group bands. When peaks are displayed in what the detector determines to be a "fingerprint" pattern, the probability of a match is high.

Raman spectroscopy results can be affected by temperature, interfering radiation, the ability to induce strong polarization of the electron cloud around a bond, and absorption of the IR frequency rather than scattering. Additionally, large molecules present more complex spectra and the resulting spectrum from an unknown material can only be compared to the finite number of entries in the associated library.

Raman spectroscopy has a large advantage over IR spectroscopy when analyzing organic components in water. IR spectroscopy detects water and the water peak can obscure other organic peaks. Raman spectroscopy is a poor detector of water and thus increases selectivity in water-based organic solutions. Raman can also analyze through most glass and translucent plastic, offering the safety of leaving the container closed during analysis (Figure 3.36).

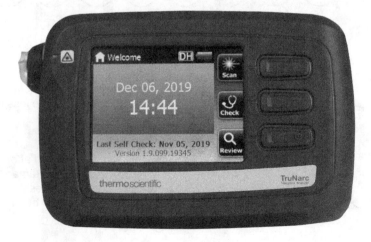

Figure 3.36 TruNarc is a hand-held Raman analyzer optimized to identify illegal drugs, precursors and cutting agents. By restricting the library database to about 400 items (compared to more than 12,000 items in the Gemini Raman library), performance is optimized for the purpose of interdicting illicit drugs. (Image: ThermoFisher Scientific, Tewksbury, MA. With permission.)

Strong Raman signal is returned from hydrocarbon compounds containing double bonds. Weak signal is returned from small, strongly polar compounds (e.g., water, methanol), most metals and elements, and dark objects that absorb laser energy. Proteins and molecules that fluoresce in the laser are poorly detected due to noise in the signal that obscures the analyte spectra.

Typically, the human eye can detect wavelengths from 380 to 700 nanometers. Hand-held Raman units for field use lasers of either 532 nm, 785 nm, or 1064 nm wavelength, with the most common being 785 nm. A 532 nm laser is visible, sometimes referred to as a green laser. The 785 nm laser is just beyond visible light, but some visible red light may travel the same path as the invisible infrared light. A 1064 nm laser is deep into the infrared spectrum.

The 532 nm laser favors analysis of inorganic materials while the 1064 nm laser favors analysis of dark samples if used properly. The 1064 nm laser produces less fluorescence, but higher heat production may occur, and scan time increases. The 785 nm is effective on at least 90% of compounds of concern. Filters, computing power, and operator technique may increase efficiency (Table 3.16).

X-Ray Fluorescence

X-ray fluorescence (XRF) spectrometers irradiate a sample with a beam of high-energy x-rays. The input of x-rays produces other, characteristic x-rays from certain elements. The change in wavelength, or fluorescence, of x-rays gives rise to the phrase "x-ray fluorescence." An XRF detector will sort characteristic wavelengths, and specific intensities are accumulated for each element.

XRF spectrometers (Figure 3.37) are used in scrap metal reclamation, jewelry validation, lead paint analysis, mining and other geological applications for elemental analysis of both metals and metal oxides. XRF spectrometers are useful in the analysis of metal, glass, and ceramic materials. Most XRF spectrometers are laboratory-based instruments. A few field portable instruments are available.

All elements absorb x-rays to some extent. Each element has a characteristic absorption spectrum that is close to the emission line of the element. Absorption attenuates the secondary x-rays leaving the sample, and the absorption coefficient can be calculated for known samples. Analysis of

Table 3.16   Relative Effects of Raman Lasers

	532 nm	785 nm	1064 nm
Sample excitation	High	Medium	Low
Fluorescence	High	Medium	Low
Heat production	Low	Medium	High

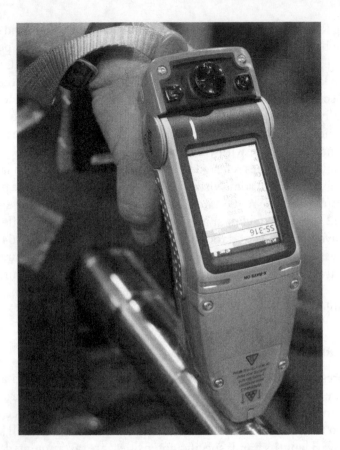

Figure 3.37 This Niton XL5 Hand-held XRF Analyzer can identify metals, metal oxides, and alloys. (Image: ThermoFisher Scientific, Tewksbury, MA. With permission.)

an unknown sample requires some presumptions that can make the analysis time consuming and complicated to most technicians. However, detection of some key metals may be of interest to some users such as those interested in screening heavy metals from hazardous waste.

Many radioactive metals may be identified with XRF after use of a survey meter. For that reason, some hand-held XRF devices include gamma detection.

An example of a field portable XRF spectrometer is the Niton XL5 Hand-held XRF Analyzer from ThermoFisher Scientific. This particular model can identify metals from atomic number 17, chlorine, through atomic number 92, uranium, as pure metals and alloys.

Lighter elements are more difficult to fluoresce and detect. Identification of atomic number 12, magnesium through atomic number 16, sulfur, is possible with the use of a silicon drift detector (SSD). SSD is a solid-state detector that enables high count rate and high resolution while geometrically positioned to maximize efficiency. The Niton XL5 uses SSD with a more powerful x-ray source to reliably identify these lighter elements in the field.

Resources

The U.S. Department of Homeland Security has information on several products related to chemical detection technology through its System Assessment and Validation for Emergency Responders (SAVER) Program. Documents are updated regularly as new information becomes available. Search at https://www.dhs.gov/science-and-technology/saver.

Radiation Detection Technology

Radiation detection technology can be used within three types of portable instruments for the purpose of characterizing hazards of unknown materials. Survey equipment has performed as the workhorse since the advent of the nuclear age. Maturing technology made portable radionuclide identifiers possible, and the more recent focus on homeland security has spurred the development of several personal alarming radiation detectors and pagers.

Integration of global positioning systems (GPS) and instrument networking complement radiation detection technologies. Hand-held instrumentation may perform functions that previously were performed by individual dedicated instruments, such as monitoring accumulated dose along with the spectrographic and radionucleide identification.

Survey Equipment

Survey equipment can be used for the detection and measurement of photon emitting radioactive substances for the purpose of detection and characterizing hazards. The most common hand-held or portable instrument technologies are overviewed below. Individual instruments may combine these technologies into a single package in order to detect a wider range of ionizing radiation.

Geiger-Mueller Tube

A Geiger-Mueller (GM) tube is filled with a gas to which a high voltage is applied when the unit is turned on. When radiation passes through the tube, an electrical pulse is generated. The pulse is modified through electrical circuitry and is output as a visual meter reading or an audible sound. This is the instrument commonly referred to as a Geiger counter.

There are three basic designs for GM probes: side window (cylindrical), end window, and pancake.

The side window design is sensitive to gamma rays. If the probe wall is thin enough, the sensor can measure high-energy beta rays (> 300 keV). A common design involves a sidewall sleeve that can be rotated to reveal a thin wall over the GM tube. Detection readings are cumulative, producing a reading that is the total of gamma and beta radiation. This design allows the operator to distinguish between gamma and beta readings because the gamma rays are always detected and the beta rays are only detected when the sleeve is rotated to expose the thin wall.

The pancake probe is a hand-held device that resembles a pancake spatula. It has a thin aluminized mylar window in front of the sensor. While it detects alpha and gamma radiation, it is most often used to detect beta source contamination.

The end window probe is a cylinder with a mica (or similar material) window at the end. It is sensitive to gamma radiation and is most often used to detect beta-emitting surface contamination. It might also be used to monitor alpha radiation.

Some models use an energy-compensated GM tube, which is useful for exposure measurements.

Scintillator

A scintillator is a material that fluoresces briefly in response to absorbing ionizing radiation. Scintillator characteristics include the amount and specific wavelength of light emitted and the duration of the fluorescence. Scintillator material will be chosen by the designer to maximize the intended function of the detector based on these and other factors (Figure 3.38).

In an operating instrument, scintillating material will emit one or more photons over a range of ionizing radiation, which affects selectivity and sensitivity. A sensor is used to detect the emission of photons, and the characteristics of the sensor must complement the scintillator. A photomultiplier

Figure 3.38 Thermo Eberline ESM FH 40 G-L radiation monitor detects alpha, beta, and gamma radiation.

tube is used to amplify the signal and pass it to a processor. A scintillating material that produces a very narrow frequency of light will make light sensing from an ionizing event more specific. Short decay time of the fluorescent flash makes the instrument more likely to detect an ionizing event that occurs immediately after fluorescence. The density of the scintillating material will affect its ability to absorb ionizing radiation; if the material is not adequately dense, it will allow radiation to pass through the scintillator undetected.

Scintillators include many materials and are matched to the application by the instrument designer (Figure 3.39). A few materials have already been shown to be most practical. Some aromatic compounds scintillate, but these are not commonly used in field applications. Scintillators containing an aromatic ring have a very fast response. When the shape of a scintillator is important, scintillating crystals dissolved in organic liquid or plastic can be shaped to meet the application.

Most survey meter scintillators consist of inorganic crystals, usually an alkali halide. The most common is sodium iodide doped with thallium and often denoted as NaI(Tl). Other scintillating materials include barium fluoride (BaF2), cesium iodide (CsI), lanthanum bromide (LaBr3), lutetium iodide (LuI3), and yttrium aluminum garnet doped with cerium (Ce:YAG). These inorganic crystals are dense and are able to stop more radiation, but the decay time of the fluorescence is longer. These materials are most effective in instruments designed to detect gamma radiation.

Other probes may be available depending on the model, such as a zinc sulfide scintillator probe for the detection of alpha radiation. Surveying for alpha radiation differs from gamma and beta surveying. Alpha emissions travel only very short distances and surveying requires deliberate, methodical movement of the probe very near an alpha source. Use of individual models is dependent on meter and sensor configuration.

Figure 3.39 Thermo Eberline ESM FH 40 G-L with enhanced scintillator for increased gamma sensitivity.

Figure 3.40 ThermoScientific TPM-903B Transportable Radiation Portal Monitor for pedestrian and vehicle use. Each post contains a large scintillator.

The TPM-903B Transportable Radiation Portal Monitor (Figure 3.40) uses a large scintillator in each post to detect as little as 1 microcurie of Cesium-137. The device is used as a mobile monitoring portal for slowly passing pedestrians and animals or vehicles if the upper arch is removed and the post separated by 10 feet. Each post contains a scintillator of about 5.2 liters with lead shielding on all but the inward side fo the post. An infrared motion detector causes the device to repeatedly sample quickly, and as contamination enters the portal an increasing radiation value causes an alarm. Residual contamination in the area will be ignored, unless grossly contaminated. Gamma and beta contamination can quickly be determined in a security event or a mass casualty event.

CZT Crystal
CZT crystal is an alloy of cadmium, zinc, and telluride that can be used in radiation detectors. Gamma radiation striking a CZT semiconductor array will generate electrons that can be detected. Certain manufacturing techniques are necessary to maximize electron detection. Uniform detector arrays with good resolution can be achieved by certain crystal doping

methods and semiconductor construction. CZT radiation detectors can be formed into required shapes for certain applications and are sometimes chosen to detect gamma radiation within a specific range.

Ionization Chamber

An ionization (ion) chamber is a gas-filled chamber that contains an anode and a cathode. When gamma radiation interacts with the sensor, an electric pulse is produced. The pulse is modified through electrical circuitry and is output as a visual meter reading or an audible sound.

High-Purity Germanium

High-purity germanium (HPGe) sensors are very sensitive and used for accurate detection of gamma and neutron radiation when a fast response is needed, such as screening of containers at a port. HPGe gamma radiation sensors have high selectivity and sensitivity and can be used to identify nuclear materials. HPGe crystals detect minute amounts of neutron and gamma ray emissions, even through heavy shielding. Some instruments can use an HPGe sensor to detect fissile nuclear materials such as uranium or plutonium.

HPGe sensors are useful in dose rates in the 0–100 mSv/hr range. Higher dose rates are typically detected with other sensors, such as a GM tube. An example of a field portable identifier with an HPGe sensor is the Ortec® Detective-Ex-100 (Advanced Measurement Technology, Inc.).

Portable Multichannel Analyzer

A portable multichannel analyzer is an electronic multichannel analyzer unit that is combined with an instrument that detects gamma radiation. When programmed with a gamma ray library and analysis scheme, gamma emitting radioisotopes can be identified by their signature.

Some ionization chambers are open to the air and ionization chamber detectors must be calibrated to a radiation source before use. They provide a direct measurement of radiation in air. Other types of ionization chambers are pressurized with air or quenched with other gases, which often contain halides.

Neutron Detectors

A neutron detector is a tube filled with boron trifluoride (BF_3) or helium-3 (He-3) and has a high voltage applied to it. When neutron radiation interacts with the gas in the tube a detectable particle is emitted; BF_3 emits a helium-4 nucleus and He-3 emits a proton. The emission pulse is modified through electrical circuitry and is output as a visual meter reading or an audible sound.

Through the use of filters and electronic circuitry, high-energy gamma, beta, and alpha radiation is rejected and the reading represents neutron radiation as measured in rem.

A multichannel analyzer is available to determine the identity of the neutron source. Neutron radiation can also be detected with a lithium iodide scintillator (LiI). Some instruments combine the LiI scintillator with another scintillating material in a single instrument capable of detecting gamma and neutron radiation. This has application in "pagers" (see two paragraphs below), among others.

Monitor Maintenance

Equipment maintenance varies by detector type, manufacturer, and model. Be sure to fully investigate maintenance and calibration requirements before incorporating a radiation monitoring scheme into your procedure for analyzing unknown substances. A detection device that is not calibrated can only provide limited information.

Personal Alarming Radiation Detectors

Personal alarming radiation detectors, sometimes called pagers, use technology similar to that used in survey instruments. Personal detectors are miniaturized and designed to wear unobtrusively by a person. They can detect radiation above a background level and alert the user. They can also be used to localize a source by detecting increasing ionizing radiation. Most are designed to detect gamma, and to a lesser extent, neutron radiation. These devices are not designed to detect alpha or beta radiation.

If you are responsible for the selection of personal alarming radiation detection equipment, be sure to read the product selection section following the radionuclide identifiers section.

Radio Isotope Identification Devices

Radio isotope identification devices (RIID) can identify hundreds of individual radionuclides. A device can measure the spectrum produced by an ionizing radiation source and compare it to a library of spectra stored in the device. Spectra are defined by decay characteristics of spin and parity, gamma peak energy, alpha and beta transitions, etc. The ability of the identifier to determine the identity of the source depends on the type and range of sensors in the device as well as the presence of a matching spectrum.

A RIID may be hand held or human portable. Human-portable RIIDs are suitcase-size units, usually in a backpack, and use a larger, more sensitive scintillator. RIIDs typically utilize sodium iodide or cesium iodide scintillators.

Over 3,500 nuclides have been cataloged. Interpretation can be complex and may not be as simple as overlaying a sample's spectra on one from the library. Sometimes a skilled interpreter is necessary.

The basic data for the library update were adopted from the NuDat 2.8 (https://www.nndc.bnl.gov/nudat2/index.jsp) database, and some corrections were made with the help of the Table of Radioactive Isotopes dated from 1999. The number of nuclides in the new library is 3386, including all

293 stable isotopes, and the number of gamma and x-ray lines is 90,223. All known deficiencies in the old library were remedied in the new one. In addition to creating a more reliable, up to date library for SHAMAN, this project contributed to correcting several errors in the NUDAT database maintained by the Brookhaven National Laboratory.

Product Selection

The U.S. Department of Energy (USDOE) National Nuclear Security Administration (NNSA) has funded an evaluation of personal alarming radiation detectors, survey equipment, and radio-isotope identifiers. *Evaluation of Preventative Radiological/Nuclear Detector Archetypes to Validate Repurpose to the Consequence Management Mission* is a resource to help you determine which detection products meet your application and need. The USDOE Brookhaven National Laboratory published the report in 2017. Public access to this report is available at, https://www.osti.gov/servlets/purl/1425190. Another resource is the *Radiation Dosimeters for Response and Recovery Market Survey Report* published by the National Urban Security Technology Laboratory, U.S. Department of Homeland Security, Science and Technology Directorate in 2016. This guide assesses commercially available equipment as tested against standards of the National Institute of Standards and Technology (NIST) and American National Standards Institute (ANSI). Access to this document is limited and is posted at the SAVER Web site, https://www.dhs.gov/science-and-technology/saver. Consult the website to determine accessibility or contact the SAVER office at 866-674-3251.

Biological Detection Technology

Several technologies are available for the detection of biological material. Immunoassay and polymerase chain reaction (PCR) methods are the most widely used. Several new or alternative methods also exist. Until recently, these technologies were motivated by analytical or diagnostic needs. As such, many technological advances have occurred in those areas, and recently the technology has been adapted to field and portable use. Additional technologies include physical and chemical properties of biological material, radioimmunoassay, optical sensors, biosensors, colorimetric, enzymatic, optical fiber with fluorescence, flow cytometry, microarray, mutual induction, luminescence, liquid chromatography, nanoparticle probes, and mass spectrometry.

Application of a particular technology depends on the setting. More specific and sensitive detection occurs in analytical and diagnostic laboratories, each having specifications unique to their mission. Portable laboratories are temporary in nature and have more interest in quick response in a disaster or other urgent need. Field identification of biological hazards is limited here to detection technology available in field portable devices. These devices must be easily transportable, able to operate in a wide range of environmental conditions, be easy to operate, have low or no power requirements, and be disposable or easily decontaminated.

No known rapid screening assay is 100% sensitive or 100% specific; therefore, results should be confirmed by an alternative method. The Centers for Disease Control requires culture of samples as the only definitive method of identification. Timely transfer to a Laboratory Response Network (LRN) facility will allow cultures to be performed and preliminary results made available within 12 to 24 hours. Your LRN can provide scalable resources for characterization and identification of suspect biological material. Field detection technology is limited to environmental sampling and should not be used as the sole basis for biological agent identification, nor should this technology be used for diagnostic purposes.

Biological detection technology can be used in screening tests and/or identifiers. A field screening test is used to generically determine if a sample could or could not be a biological agent. An identifier is a test that determines the specific identity of a biological agent. Test results are based on a combination of specificity and sensitivity to a particular target. Some tests may be used to screen, identify, or do both. Some tests are portable, other tests are laboratory based, and some are adaptable to field use.

Screening and Identifying Technology

Hand-held test kits have been used for years by military personnel and first responders under narrow response criteria. In general, they are used when a specific threat is identified through a hazard assessment. For example, the DoD BSK hand-held test was used to presumptively identify powder found in a letter to Senator Daschle in Washington, DC, during 2001 as *Bacillus anthracis.*

Identifying technologies are capable of identifying the species of cellular and viral agents as well as toxins. (The United States denotes toxins produced by living organisms to be a biological agent, even though the toxins are not alive or capable of reproducing; other nations group these toxins with chemical agents.) Tests using identifying technology detect the presence of one or more unique biomarkers characteristic of a suspect biological agent. Identifying tests are based on two main categories: antibody-based tests and gene-based tests. Antibody-based identifiers are quick and prone to automation. Gene-based identifiers are more exacting, but require more effort and time.

When selecting a technology and system for biological agent detection, it is important that both sensitivity and specificity of a particular test be measured relative to a reference standard. Sensitivity and specificity are jointly determined at the threshold for a positive test, because either can be made excessive at the expense of the other. Considering sensitivity or specificity alone is not useful.

Physical and Chemical Properties

Hand-held chemical screening tests can be used to rule out biological material. A set of assumptions is used to determine chemical and physical parameters of an unknown material. If the material is a visible solid or liquid, constitutes a volume large enough to sample for a few simple tests and appears to be relatively pure, the operator can rule out some chemical and physical properties of biological material.

Screening tests are not conclusive. They must be used in conjunction with a hazard assessment unique to each situation. Physical and chemical properties of biological warfare (BW) agents in bulk and some simple tests and observations are listed below.

Appearance

Biological agents are solids or liquids. Finely divided powder can indicate the material was manipulated to improve dispersion in air (BW agent) or water (organic pesticide). The material is light tan to black in color due to the organism coating or residue from growth media. Brilliant white (powdered sugar, baking soda, table salt) is a clue that the material is not a biological material.

Biological agents may be present in a volume too small to see, a situation that falls outside the scope of this book. In cases where a small volume can be visualized, care should be taken not to destroy the entire volume in testing. The sample should be split with enough preserved for additional testing and evidence (Figure 3.41).

pH

pH for biological material must be relatively neutral for the organism to remain viable. pH of less than 4 is too acidic and pH greater than 10 is too basic for organism survival. A pH of 4–10 means the material could be a biological agent.

Protein

The presence of protein is common to all biological agents and toxins. A protein test with a detection limit of 1% is sufficient to test bulk biological material. If a biological agent is suspected to be hidden in a non-BW agent material, a more sensitive and specific test would be needed, so results of a protein test need to be considered in the context of the hazard analysis.

A protein screening test should test for protein generally and not test for a specific protein. It should also operate throughout the pH range of biological material. Protein is known to reduce copper sulfate to produce a blue to

Figure 3.41 Split the sample so that the entire volume is not consumed in testing. Preserve some of the material for future testing and some as possible evidence. If testing cannot be performed on a very small amount, you should reevaluate your characterization strategy.

purple color change, which can be difficult to visualize in dilute samples. All of the following methods are suppressed in the presence of strong reducers.

The biuret method of protein detection uses biuret reagent, a strongly caustic solution of copper sulfate. This deep blue solution can be diluted with water in a test tube. A small sample may be added to the test tube or the dilute solution may be added directly to a powder on a surface. If protein is present, a purple color (540 nm) will develop from blue. The color change is difficult to see in the case of low concentrations of protein. The result is sensitive to the presence of nitrogen-containing compounds.

The Bradford method causes a complex of protein with Coomassie Brilliant Blue G-250 dye in phosphoric acid and methanol. The protein-dye complex causes a shift in the absorption maximum of the dye from red to blue (465 to 595 nm). The amount of absorption is proportional to the protein present.

The Lowry method combines sodium carbonate, sodium potassium tartrate, cupric sulfate, and phosphomolybdotungstate in water. When mixed with protein, a color change to an intense blue-green is observed.

Copper sulfate combined with bicinchoninic acid (BCA) produces a clear, apple-green solution. Protein will complex with BCA and copper to cause a color change to purple. The color change is easily discernable visually with protein concentration as low as 1%. This solution is the simplest to mix, is stable, and may be sprayed directly on a suspected protein or applied to a white absorbent test bed of paper or cotton (Figure 3.42).

Protein test strips used for urinalysis are not dependable for field detection of protein. They are often designed for a narrow diagnostic use, such as

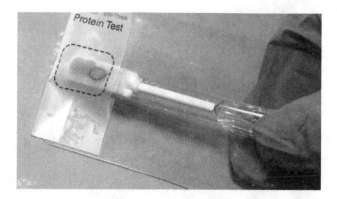

Figure 3.42 A simple protein test is a useful screen for biological material. This test uses copper sulfate on a polyethylene test bed. The powder is collected with tape, folded onto the test bed, and a capsule of bicinchoninic acid (BCA) is crushed. The BCA and copper produce a bright green. The presence of protein produces a dark purple complex with BCA and copper.

the detection of a certain protein in urine. As such, they may not be able to function across a wider range of pH or detect protein universally.

Starch

A reliable starch test can confirm the presence of starch. Starch is a component of plants. While starch does not rule out biological material, it does point to food products. Starch is known to degrade the dispersion capability of a biological warfare agent. Application of iodine to starch causes the dark brown iodine solution to turn black.

A solution of about 0.5% iodine in water stabilized with potassium iodide may last a year in an amber glass bottle. Povidone (Betadine®) is a better choice for long shelf life. Povidone has about ten times the available iodine and has a five-year shelf life. At full strength, the brown color is so strong that the color change to black is difficult to see. Povidone may be applied full strength or diluted 10 or 20 to 1. Testing on a white background such as absorbent paper or cotton will help wick the excess brown liquid away and expose the blackened starch (Figure 3.43 and Figure 3.44).

Milk Products

Milk products can be detected if sugar is not present by the use of an enzyme, beta-galactosidase (Lactaid®). Milk contains the carbohydrate lactose. Lactase will break most of the lactose in a small sample into glucose and galactose within a few seconds. The increase in glucose can be detected by a glucose urine test strip (Bayer® Uristix™ Glucose Strips). If the sample already contains a high amount of glucose (flavored milk) before the test, the test strip

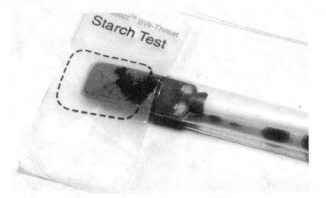

Figure 3.43 A simple starch test indicates the presence of starch, a common food product. A dilute povidone solution provides free iodine, which produces a dark black stain in the presence of starch.

Figure 3.44 Filtering may be useful to concentrate a powder for testing. Here a dilute povidone solution is poured or sprayed over flour on the filter.

will be overloaded and the small increase in glucose from the lactase will not be detected. Dipping a glucose strip in the sample before the addition of the lactase will provide a benchmark for the glucose strip used after the addition of lactase. The test can be performed in less than 30 seconds. This test for milk is illustrated in Figure 3.45 through Figure 3.48.

The presence of milk must be considered in conjunction with the hazard assessment because powdered milk could be harmlessly confused with a biological agent; however, milk can also be an ideal media for nurturing biological material.

Plant-Based Carbohydrate

Many plants contain melibiose, a complex sugar that causes gastric distress in some people. Melibiose is present in legumes, such as peas, beans, soybeans, and lentils, all of which may be dried and ground into food powders.

Figure 3.45 Beta-galactosidase (Lactaid) can be used in small amounts to produce glucose, which can be detected with a urine glucose test strip.

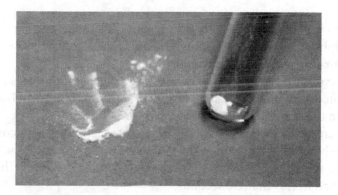

Figures 3.46 A small amount of suspect milk product is added to a test tube.

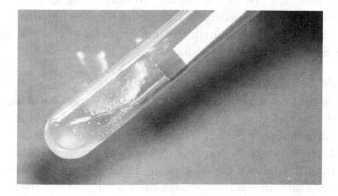

Figure 3.47 A small amount of water is added in the next step, followed by a glucose test strip.

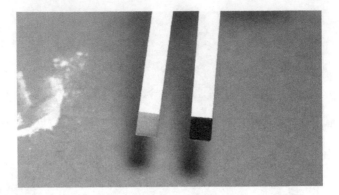

Figure 3.48 If the solution does not contain glucose, a small amount of powdered beta-galactosidase is added and mixed for a few seconds. A second glucose test strip is used to determine a relative increase in glucose; negative result is on left and positive result is on right. An increase in glucose indicates the presence of lactose.

Melibiose will break down to glucose and galactose when exposed to the enzyme alpha-galactosidase (Beano®). Alpha-galactosidase is the stereoisomer (mirror image) of beta-galactosidase.

The increase in glucose can be detected by a glucose urine test strip (Bayer Uristix Glucose Strips). If the sample already contains a high amount of glucose before the test, the test strip will be overloaded and the small increase in glucose from the melibiose will not be detected. Dipping a glucose strip in the sample before the addition of the melibiose will provide a benchmark for the glucose strip used after the addition of alpha-galactosidase. The test can be performed in less than 30 seconds.

The presence of melibiose suggests a powder may be of plant origin; however, melibiose could be part of a fermentation mixture for a BW agent. It is unlikely that a free sugar would be mixed with a deployed BW agent since the sugar would absorb moisture and reduce mobility of the agent.

These two tests show, generally, that an increase in glucose from alpha-galactosidase indicates plant material; an increase in glucose from beta-galactosidase indicates a dairy product.

Hand-Held Sampling Device

If a bulk amount of suspected biological sample is not available and wipe sampling of surfaces is not an option, a hand-held sampling device can be used in conjunction with biological test kits that require a small sample. The sampling device is a way to concentrate biological material from the air by collecting the material in a filter cartridge. The filter media in aqueous solution produces a sample that may contain enough suspect material to be tested by various tests. The solution may be used with any liquid-based sensor system, although the specific test may require preparation.

Generally, a hand-held sampling device will not collect enough material to process with the tests mentioned above and must be processed with one or more of the technologies to follow.

Immunoassay

Immunoassay technologies detect and measure the highly specific binding of antibodies with their corresponding antigens by forming a three-dimensional, antigen-antibody complex. Overall, they are quick and accurate tests that can be used to detect specific molecules.

An antibody (or immunoglobulin) is a Y-shaped protein used by many organisms to defend against bacteria, viruses, and other material. An antibody recognizes a specific foreign material, called an antigen, by a specific bonding site, which only "fit" the antigen, located at the ends of the Y-shaped protein. Each antibody has a unique structure that attaches to an antigen in a lock and key fit. When the antibody is attached to the antigen, the antigen is destroyed or marked for destruction or elimination by some other method.

An antigen is a substance that stimulates a response from the immune system, usually the production of antibodies. Antigens are usually proteins or complex sugars, but may be other material.

An immunoassay based test is highly specific to an antigen in a sample by the action of its corresponding antibody. The two react to form an antigen-antibody complex, which triggers some readable result such as a color change.

Antibody-based tests can be highly specific and sensitive. Detection thresholds can be as low as 1,000–10,000 microbial cells per milliliter. Disadvantages of antibody-based tests include false-positive results due to nonspecific binding, known as the hook effect, and cross reactivity, known as the matrix effect. An accepted method of overcoming potential false-positive results in antibody-based bioassays is to run each assay in triplicate in sample dilutions of 1:1, 1:10, and 1:100. False-negative results can be caused by degradation of the antibodies over time, which also determines shelf life.

Immunoassays are typically used in one of four formats: monoclonal-polyclonal sandwich, antigen-down, competitive inhibition, and rapid.

Monoclonal-Polyclonal Sandwich Immunoassay

Monoclonal-polyclonal sandwich immunoassay uses a monoclonal antibody (sometimes called a capture antibody) fixed to a plate (Figure 3.49). The sample is added and if the corresponding antigen (analyte) is present, it attaches to the antibody and is immobilized. A polyclonal antibody (sometimes called a detecting antibody) is added, which also binds to the antigen. The two antibodies have formed a "sandwich" around the antigen. The plate is rinsed and only attached, detecting antibody remains. The detecting

Monoclonal or capture anti-bodies are fixed to a plate.

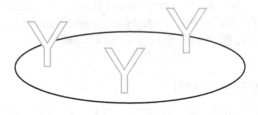

The sample is added and if the antigen is present, it attaches to the antibody and is immobilized.

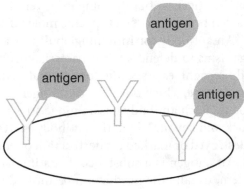

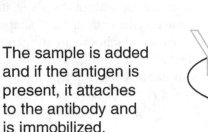

A marked antibody is added, which attaches to the antigen-specific antibody. After rinsing, signal is produced by the marked antibody.

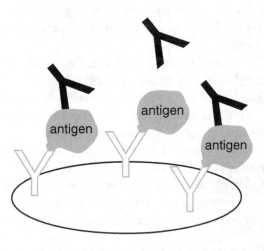

Figure 3.49 Monoclonal-Polyclonal Sandwich Immunoassay.

antibody can be measured by various methods, but the signal is proportional to the concentration.

Enzyme-Linked Immunosorbent Assay (ELISA) uses an enzyme step to amplify results (Figure 3.50). ELISA is a sensitive biochemical technique used to detect the presence of specific substances, such as enzymes, viruses, or bacteria in a sample. ELISA is based on the concept of antigen-antibody complex

ELISA uses an antibody
bound to a plate, which
attaches to an antigen from
the sample. A second
antibody forms a sand-
wich, similar to the
monoclonal-polyclonal
immunoassay.

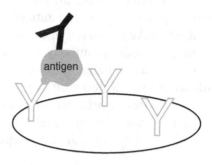

ELISA methodology
can amplify the signal from
low amounts of antigen.
A third antibody with an
enzyme is added and
attaches to the second
antibody and then the
plate is rinsed.

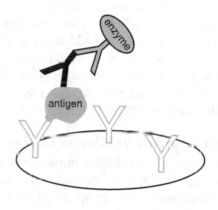

A substance is added
which the enzyme can
rapidly convert to a
visible dye or other
signal. Only a few
bound enzyme mole-
cules can indicate the
presence of the antigen.

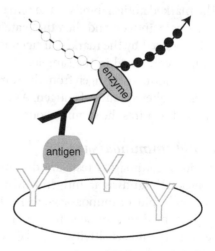

Figure 3.50 Schematic representation of an ELISA.

with two antibodies and an indicator, usually an enzyme and a dye. The first antibody is specific to the antigen in the sample. The indicator is another antibody linked to an enzyme that reacts with the antigen-antibody complex or an intermediate antibody. The antigen-antibody/antibody-enzyme complex produces a change in color or fluorescence from a previously invisible substance to indicate a positive result. The enzyme acts as an amplifier to the color or fluorescence because only a few bound complexes will produce many signal molecules through catalytic action. There are minor variations on this procedure. ELISA tests are time dependent.

Antigen-Down Immunoassay

Antigen-down immunoassay, also called direct immunoassay, detects antibodies that might be present in a sample (Figure 3.51). An antigen is bound to a plate. The sample is added to the plate and if the target antibody is present in the sample, it binds with the antigen and is immobilized on the plate. A second antibody is added that binds with the first antibody. The second antibody is or can be tagged with a dye to produce a colored or fluorescent signal that may be detected visually or with a reader.

Competitive Inhibition Immunoassay

Competitive inhibition immunoassay can be used for small, easily bound antigens. A specific antibody is bound to a plate (Figure 3.52). The sample is added, and if the corresponding antibody is present it binds to the antigen and is immobilized on the plate. Next, a known amount of marked antigen is added. The marked antigen binds to the remaining antibody sites not already occupied by sample antigen and then the plate is rinsed. The color change or other signal produced by the marked antigen is inversely proportional to the amount of antigen contained in the sample – that is, strong color development indicates lower amounts of antigen from the sample while lack of color development indicates a higher amount of antigen. Another competitive inhibition immunoassay method requires the simultaneous addition of sample and marked antigen.

Rapid Immunoassay

Rapid immunoassay tests are designed using combinations of monoclonal-polyclonal sandwich, antigen-down, and competitive inhibition immunoassays. Rapid immunoassays were developed primarily to fill the need for a quick field test for specific material that retained an acceptable degree of selectivity and sensitivity. Most often, the result is expressed through development of a color and more specifically these are called immunochromatographic tests.

Hand-held immunochromatographic assays are single-use devices that are often compared to a home pregnancy test kit. They provide a "yes or no" result based on the formation of two colored bands. One band is a control to

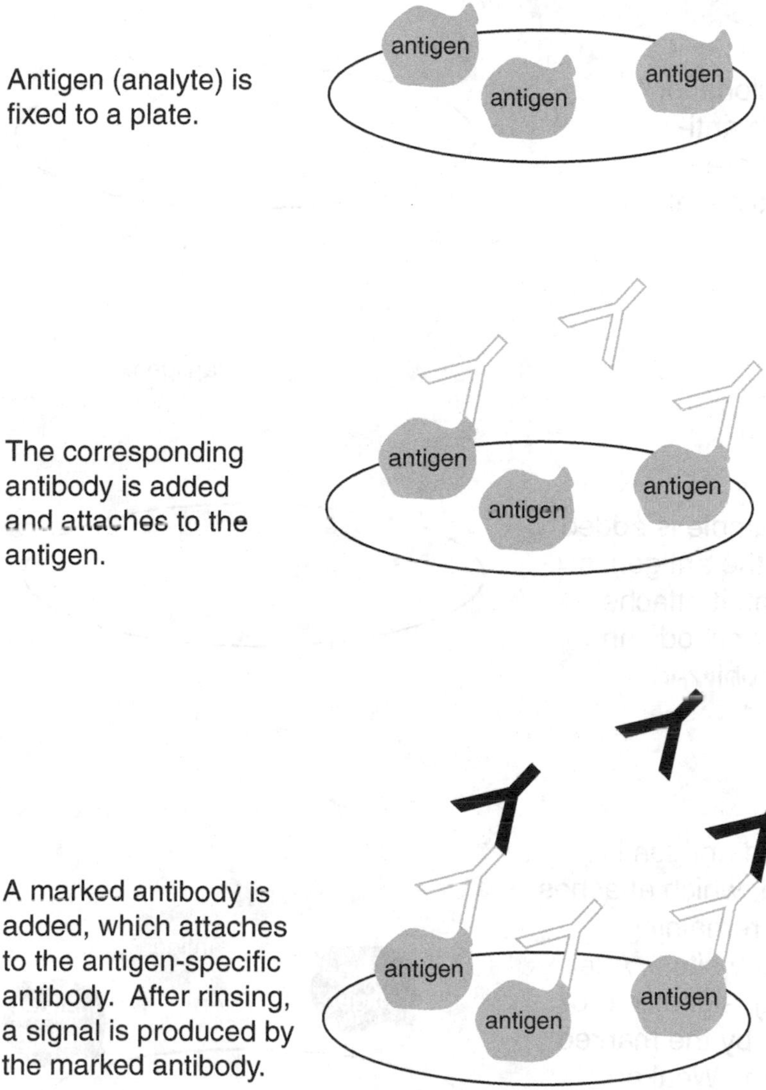
Antigen (analyte) is fixed to a plate.
antigen
antigen
antigen
The corresponding antibody is added and attaches to the antigen.
antigen
antigen
antigen
A marked antibody is added, which attaches to the antigen-specific antibody. After rinsing, a signal is produced by the marked antibody.
antigen
antigen
antigen

Figure 3.51 Schematic representation of an antigen-down immunoassay.

assure the test is working; the second band represents a positive result for a specific agent. The test is designed to be qualitative; however, the degree of color change in the second band indicates a semiquantitative measurement. The use of an optical reader can estimate the concentration of biological agent present in the sample (Figure 3.53). These tests typically must be read after 15 minutes and before 30 minutes or results are not valid.

Lateral flow immunochromatography uses monoclonal antibodies that are specifically attracted to the target substance. This technique reduces

Monoclonal or
capture anti-
bodies are
fixed to a plate.

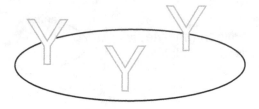

The sample is added
and if the antigen is
present, it attaches
to the antibody and
is immobilized.

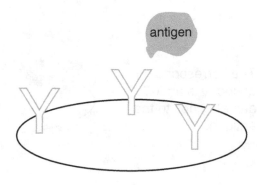

Marked antigen is
added, which attaches
to the remaining
antibody sites. After
rinsing, signal is pro-
duced by the marked
antigen. Weaker
signal (color) indicates
more antigen from
the sample.

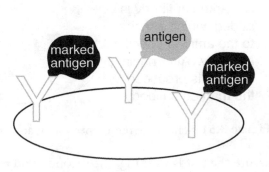

Figure 3.52 Schematic representation of a competitive inhibition immunoassay.

false-positive results in environmentally collected samples. Lateral flow immunochromatography tests can be specific and sensitive, but not always sensitive
enough to be considered definitive. A concentration at the detection limit of
the test can be higher than the amount that can cause infection. An example of
the lateral flow format is the BADD™ Anthrax Rapid Detection Device.

Figure 3.53 The Guardian Bio Threat Alert, a hand-held immunochromatography test with optical reader.

The test is performed by placing a small amount of prepared liquid sample directly on a filter strip. Alternately, a solid can be collected with an extract solution or a solid can be placed on the strip and moistened with a recommended volume of buffer solution. Most tests are volume dependent; too much liquid floods the device and too little prevents the movement of material through the test strip. A signal reagent specific to the analyte is embedded in the filter strip and dissolves into the liquid. The signal reagent binds to the antigen or antibody and moves across the filter strip by capillary action.

A second antibody or antigen is immobilized on the filter strip at the point where a colored band would appear. The second antibody is combined with a dye, colloidal gold, or other conjugate to generate signal. As the analyte-signal reagent moves through the region of this second antibody or antigen, the complex is captured, concentrated as liquid moves through it, and forms a color. Excess antibody flows to a control area and forms a color in contact with immobilized material to prove the test is functioning correctly.

If the level of the target substance is present in the sample above a certain concentration, the antibodies and target substance combine to form a colored band.

Samples containing a higher concentration of antigen (agent) than marked antibodies can cause a false-negative, which is sometimes referred to as the "hook effect." In this case, a high proportion of agent remains unmarked compared to the relatively few that bond to all available marked antibodies. The unmarked antigen is able to occupy bonding sites in a competitive manner and prevents the marked antibodies from concentrating to a visible amount in the test result strip. Hook effect usually occurs in tests involving toxins rather than organisms.

Hook effect can be eliminated by diluting the sample at a ratio of approximately 100:1 and 10,000:1 in buffer solution. To prepare the solution place two drops of the originally prepared buffered solution from the first test into a test tube and mix with 10 ml of new buffer to form the 100:1 solution. Use this solution in a new immunochromatographic assay. Place 2 drops from the 100:1 buffer solution into another test tube and dilute that with 10 ml of fresh buffer to form the 10,000:1 solution. Use this solution in a new immunochromatographic assay. If your buffer supply is lacking, you can perform the dilutions by adding one drop to 5 ml of buffer.

By performing tests on all three dilutions you have covered a range of dilutions in which at least one dilution will be optimized for detection by a hand-held immunochromatographic assay and you will have eliminated the hook effect.

Another type of hand-held immunochromatographic assay can detect and identify multiple analytes. For example, Sensitive Membrane Antigen Rapid Test (SMART) detects antigens by immuno-focusing colloidal gold-labeled antibody reagents and their corresponding antigens onto small membranes. The gold complex forms a red spot or band that can be detected by the operator. Excess antibody flows to a control area and forms a color in contact with immobilized material to prove the test is functioning correctly.

Immunochromatographic assay false-positive results can occur due to nonspecific binding of the antibodies. False-negative results can occur due to degradation of the antibodies over time. Tests can only be created based on limited availability of antibodies.

Hand-held immunochromatographic assay tests, sometimes called tickets or kits, are disposable matrix devices. They are normally stored in a dry and cool location to avoid degradation of the biological or other reagents in the test. The tests are opened at the point of use and run wet when a liquid sample or when water, usually in the form of a buffer solution, is added.

Rapid hand-held assays with greater sensitivity, specificity, and reproducibility are emerging as the technology matures. They are for use on a wide range of bacterial agents and toxins. These assays are expected to have stable and long shelf life and will be easier to operate. Hand-held assays are field portable, have no power requirement, and are easy to use. Simple sample preparation and data interpretation make them preferable over gene-based assays for rapid field detection. However, hand-held assays are prone to false-positives.

ANP Technologies, Inc. has developed a hand-held assay device that reduces false-positives while detecting the presence of biological agents at extremely low concentrations. The device has been engineered with proprietary nanomanipulation technique that aligns the antibodies so that they are more available for reaction. Reportedly, this enhances assay sensitivity up to 100-fold with smaller sample volume. A protein array microdesign allows simultaneous testing for several biological organisms and toxins.

Nucleic Acid Amplification

Nucleic acid amplification, also known as polymerase chain reaction (or PCR), can replicate many copies of DNA from minute, previously undetectable amounts within a sample.

Polymerase chain reaction technology uses a polymerase, a naturally occurring enzyme, to catalyze the formation of DNA or a characteristic segment of DNA. DNA is formed from two complementary, not identical, chains of amino acid, which are bound together. If the DNA is torn in half (down the center, like a zipper), polymerase can recreate the complimentary chain missing from each half, which results in two strands of DNA identical to the beginning chain (Figure 3.54).

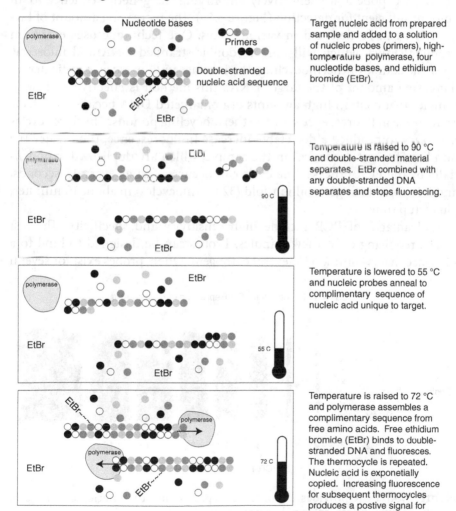

Figure 3.54 Schematic representation of PCR.

The DNA of interest is supplied by the sample and becomes the target sequence of genetic material for the test. The complementary strands of any DNA present in the sample are separated by heating. Synthetic DNA segments known as nucleic acid probes are complementary to the sequence at an end of each of the target strands. The synthetic DNA segments attach to the single-strand DNA and serve as primers for polymerase. Only genetic material that is bound to the nucleic acid probe can be replicated in the test. The polymerase enzyme begins at the primer and replicates the strand from amino acids added to the test solution.

Cooling the solution allows double-stranded DNA to remain intact. Subsequent heating and cooling cycles (thermocycling) will replicate the DNA exponentially, effectively amplifying the genetic sequence to an amount sufficient for detection (Figure 3.55). The increasing amount of target DNA can be detected in various ways. One technique uses ethidium bromide (EtBr), which will bind to double-stranded DNA and fluoresce. Over the course of thermocycling, the increase in fluorescence indicates an increasing amount of the target DNA. This method is effective even with samples that contain high amounts of nontargeted DNA because it detects an increase in fluorescence over the thermocycling periods. The fewer cycles necessary to produce a detectable increase in fluorescence, the greater the number of target sequences in the sample, which are duplicated exponentially by the PCR process. The cycle can occur as fast as every 17 seconds, and it is possible to get a billion-fold (32 thermocycles) in about 10 minutes, plus setup time.

Advantages of PCR include high sensitivity and specificity. The test is fast, requiring only a few minutes. Probes can be designed to bind to a sequence that is common to several pathogens. Often, probes exist for several

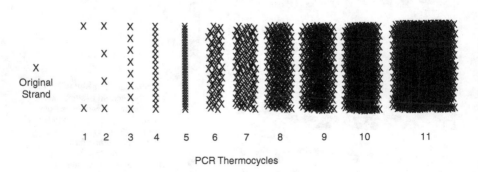

Illustration of Exponential PCR Replication of Nucleic Acid

PCR Thermocycles

Figure 3.55 PCR provides exponential reproduction of genetic sequences to rapidly produce a detectable volume from the sample.

genetic aspects of each organism to identify specific species or origins of threat agents (Anthrax Sterne strain (34F2) versus Ames strain).

Disadvantages of PCR include the need for pathogen-specific probes, which may not be available. Additionally, a specific probe must be used for each pathogen, making PCR screening tests time consuming. Depending on the purity or refinement, PCR may not work on toxins because a suspected nucleic acid sequence may not remain in the aliquot.

An example of a portable PCR device is the Advanced Nucleic Acid Analyzer (ANAA) or Mini-PCR developed by Lawrence Livermore National Laboratory. This device has been peer-reviewed. Belgrader et al. report the ANAA was able to detect 500 CFUs of *E. herbicola*, a vegetative bacterium used as a surrogate for *Y. pestis*, in 15 minutes (about 25 thermocycles). After thermocycling modifications, the ANAA was able to detect 500 CFUs of *E. herbicola* in 7 minutes. Ongoing software and automation improvements are intended to reduce the time required to run the tests and to protect the operator. The ANAA utilizes 10 silicon reaction chambers with thin-film resistive heaters and solid-state optics. The ANAA offers real-time monitoring, low power requirements for battery operation, and no moving parts for reliability and ruggedness.

Table 3.16 summarizes performance characteristics of immunoassays, PCR, and culture for the detection of *B. anthracis*.

Radioimmunoassay

Radioimmunoassay (RIA) is a highly sensitive laboratory technique used to measure minute amounts of substances. It involves marking an analyte with a radioactive marker and then determining the presence and quantity of the analyte by detection of radioactivity. In this way, an antibody could be coupled to an antigen that contained any one of a number of radioactive atoms. Many medical tests that utilized this principle have been replaced by nonradioactive tests, but the technology can still be applied to detection and quantification of biological material as well as other substances to which an antibody can be attached (Table 3.17).

Because radioimmunoassay uses radioactive material as tags, it is difficult to utilize in field applications.

Optical Sensors

Optical-sensing technology can detect biological material as broad classes of microorganisms and not as individual agents. The detector is an optical transducer that detects fluorescent or other marker molecules tagged to agents of interest. A multisensor array may be used to interrogate a sample

Table 3.17 Performance Characteristics of Bacterial Culture, PCR, and Hand-Held Immunoassays for the Detection of *B. anthracis*

	Microbiology/Culture	Polymerase Chain Reaction (PCR)	Hand-Held Immunoassays
Minimum limit of detection (spores)	1	100–1000	100,000 to 100 million
Assesses viability	Yes	No	No
Nonspecificity	No	Yes (near-neighbor bacteria)	Yes (near-neighbor bacteria and chemicals)
Other issues	Culture may require days to complete	Technology immature	Susceptible to interferences

Source: Table adapted from *Update on Biodetection: Problems and Prospects*, U.S. Department of Health & Human Services.

and discriminate classes of microorganisms. In this manner, biological agents can be distinguished separately from naturally occurring microorganisms. An optical sensor system can be used as a screening method.

Biosensors

Biosensors use information from biological materials to detect substances (Figure 3.56). Biosensors include organisms, bacteria, or proteins sensitive to a material of interest. The canary in the coal mine was a biosensor, but it was slow to respond and not very sensitive; the warning came after the miners had already been exposed to harmful levels of toxins.

Modern biosensors examine the response of the biosensor in greater detail and specificity. Typically, these sensors have been primarily biological themselves, consisting of proteins and bacteria known to react with the desired analyte. Their biological composition made them inherently slow, expensive, and difficult to maintain. Current solid-state biosensors utilize fiber optic and planar waveguides to measure changes in the properties of light and evanescent waves an analyte has bonded to its surface. Surface plasmon resonance, grating couplers, and resonant mirrors are also being used in biosensors.

Colorimetric

Colorimetric technology may be useful in the detection of volatile compounds produced by biological agents. Some colorimetric tests are capable of sensitivity in the ppb (parts per billion) range. When several different colorimetric tests are used simultaneously, the specific volatile compounds

Figure 3.56 Modern biosensors are much more sophisticated than this, but don't overlook the obvious when collecting site information that can help you characterize the hazards of the unknown material.

produced by microorganisms can be arranged in a matrix and compared to a reference.

The technology is not yet fully mature for biological detection application.

Enzymatic

Enzymatic technology uses enzymes to manipulate and sort genetic material. With the addition of several enzymes and several steps, certain genetic structures are recognized if present and then cleaved, paired, and tagged.

With proper temperature maintenance, enzymes can cycle on the target organism DNA repeatedly, thus amplifying the result. A fluorescent or colorimetric tag is applied to the replications to produce a signal and indicate a positive result.

Several manipulations are necessary and training is required.

Optical Fiber with Fluorescence

Optical fiber technology can detect some biological agents without fluorescence; others require fluorescence for detection. Light conducted through the optical fibers will scatter light out of the fiber at certain wavelengths based in part on the refractive index of the surrounding material. Optical fibers are coated with a film that captures a target microorganism. The presence of the microorganism will alter the light being scattered by the fiber. Some microorganisms can be detected by specific light-scattering characteristics. Others can be detected by use of a fluorescent tag.

Optical fiber detection of biological agents is an immature technology that could be useful in field applications. Results can be obtained in about 10 minutes, but training is not intuitive.

Flow Cytometry

Cytometry is the measurement of physical and chemical characteristics of individual cells. Flow cytometry technology measures the characteristics of the particles in a stream of fluid with laser light-scattering detection and with dyes bound to certain parts of the cell that fluoresce. Properties of scattered laser light can be interpreted to indicate the size and number of particles in the stream of fluid as well as the amount of DNA, presence of specific nucleotide sequences, and cellular proteins.

This technology utilizes the dynamics of the flowing stream of a particle-containing fluid combined with laser optics, a detector, signal processor, and computer to produce a result in a few seconds. Throughput is on the order of thousands of cells, but mixed material can be treated with a fluorescent dye that binds to biological material. Flow cytometry is reported to be 200,000 times more sensitive than gel electrophoresis and able to produce results in about 10 minutes.

Flow cytometry instrumentation has been available since the 1970s. A more recent example of available flow cytometry equipment is an instrument developed by Los Alamos National Laboratory (LANL) named the Flow Cytometer, also known as the "Mini-Flow Cytometer." It uses a helium-neon laser diode and two laser light detectors to measure scattering and particle size. Two photomultiplier tubes detect fluorescence from dye bound to biological material. The Mini-Flow Cytometer weighs about 30 pounds and is typically a benchtop instrument. It could conceivably be adapted to field use.

Smaller and more portable field equipment is being developed for detection of biological material. These newer units have the advantage of additional laser/detector combinations, filters, and dyes for wider and more specific detection capability. Operation of flow cytometry instruments requires specialized training and knowledge that reduces practicality for field use. This technology is more likely to be utilized in a laboratory when a sample has been transported there for analysis.

Microarray

Microarray technology is used to identify biological agents by determining genetic information. A microarray is a collection of microscopic spots attached to a surface. The spots contain fragments of genetic material known as probes. The sample material is prepared and exposed to the probes. The probes bind corresponding portions of genetic material and the result forms a "map" on the microarray. A sample can be matched to the genetic fragments on the array to produce a corresponding pattern.

The microarray can be used for gene expression or DNA sequencing. A sampled sequence can be matched to a known sequence for several biological agents. This technology is useful in analytical and diagnostic applications, but does not appear to be currently adaptable to field use.

Mutual Induction

Mutual induction technology is a method of detecting specific biological agents magnetically. It uses magnetic markers bound to a target material in a manner that maximizes saturation of the markers around the analyte. The sample is placed in an electromagnetic field and detection occurs through mutual induction. Training and use are simple. Results are produced in about 20 minutes.

Luminescence

Luminescence technology detects specific bimolecular material by light emission. A target substance is captured on a surface and characteristic light is produced when the complex is subjected to an electrical field, chemical reaction, or a beam of light from a diode or filtered lamp. This integrated technology is developing for field use with some systems in military use.

Liquid Chromatography

Denaturing high-performance liquid chromatography was developed to detect mutations in genetic material by reading DNA sequence. Sample preparation requires PCR, which makes it impractical for typical field use. This technology has better application in an analytical setting.

Nanoparticle Probe

Nanoparticle probe technology directly measures DNA, RNA, and protein targets without amplification (PCR). Specific agents are detected by using target-specific oligonucleotides conjugated to gold nanoparticles. The target is captured between the nanoparticle probe and a capture strand embedded in a chip. Light scattering is measured through an optics system and signal characteristics are processed by computer.

Nanoparticle probe technology is still maturing. Samples must be prepared and analysis takes about an hour.

Mass Spectrometry

Mass spectrometry uses as little as a few nanograms of sample to identify molecular weight and structure characteristics. The sample, through various methods, is introduced to the detector in a flow of air or gas. The sample is ionized by an energy source and the ions move toward detection surfaces. The mass spectrum that is produced based on drift time, etc., is compared to a library to determine if the sample is consistent with mass spectra for biological material.

Edgewood Chemical Biological Center (ECBC) and the University of Utah jointly developed the Pyrolysis-Gas Chromatography-Ion Mobility Spectrometer (Py-GC-IMS), which burns the biological particles. These pyrolysis products are then separated using gas chromatography. Columns of separated pyrolysis products are moved into an ion mobility spectrometer for analysis. A computer compares the mass spectrum to a library and produces a result. This process completely destroys the sample and detects atoms and biomolecular fragments.

The system is designed to perform trigger, detector, and classification functions. It can identify chemical aerosols and vapors in pure form and in mixtures. It can be used to characterize biological aerosols, but not identify them.

The Matrix-Assisted Laser Desorption Ionization-Time-of-Flight-Mass Spectrometry (MALDI-TOF-MS) is a variation of mass spectrometry that uses "soft ionization" of the sample (Figure 3.57). This more gentle approach is less destructive than pyrolysis and can be used to identify a biological material, even in mixtures, by producing mass spectra characteristic of individual biological agents due to limited fragmentation of sample material.

Biological material can be identified if the sample is relatively pure through a process called mass-mapping. This process requires breaking biological molecules, such as protein, into specific fragments through treatment with specific enzymes. Thus, specific proteins can be determined by the mass spectra produced by the total of its components.

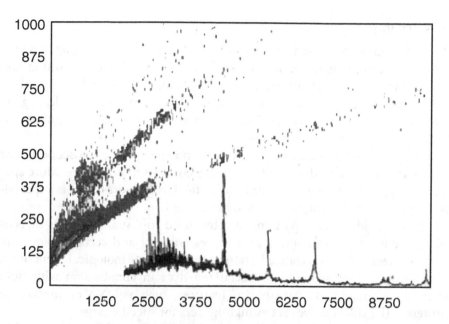

Figure 3.57 Example of a theoretical ion mobility MALDI mass spectrum with UV MALDI mass spectrum shown at bottom.

This technology is capable of high-throughput once the sample has been prepared. Operation of MALDI-TOF-MS instruments and the associated sample preparation requires specialized training and knowledge, which reduces practicality for field use. This technology is more likely to be utilized in a laboratory when a sample has been transported there for analysis. Development of automated field portable MALDI-TOF-MS is expected.

Mass spectrometry of proteins is an emerging method for the characterization of proteins. Proteins are ionized mainly by electrospray ionization or MALDI technique. In some cases, the protein can be directly introduced to the mass spectrometer; in other cases, the proteins are first broken into peptides using enzymes that cleave proteins at known locations. MS detection of proteins or peptides requires mainly laboratory-based equipment as MS detection occurs through time-of-flight, Fourier transform ion cyclotron resonance, quadrupole ion trap, or MALDI techniques. The detector is subject to "flooding" if too much sample is introduced because the ionization source may not be able to ionize the analyte completely; this results in a "muddy" spectrum. Additionally, impure samples can be difficult to discern due to the presence of unanticipated biological material. Unfortunately, the less expensive the technique, the more sample preparation is involved.

A number of different algorithmic approaches have been described to identify peptides and proteins with mass spectrometry, all of which are laboratory based. As this technology matures, field application may emerge.

Microscopy

A microscope can be used to determine visual characteristics that can provide clues to the characteristics of an unknown solid. Some training and practice will allow you to differentiate gross biological material from nonbiological material. The method does not specifically identify a biological agent, nor does it guarantee a sample does not contain some small amount of biological agent; it is merely suggestive.

The bright field microscope is the best known type of microscope. Other optics are included in dark field and phase contrast microscopes. More specialized types exist but are not common to field use. Fluorescence is a specialized application that might be useful in the field.

Bright field microscopy is most often used to visualize spores or cells that have been stained. Staining increases contrast and certain colors can be visualized when specialized stains are bound to biological structures. Fluorescence microscopy uses specialized dyes in conjunction with light altering techniques to mark specific features, such as DNA or specialized proteins. The colorful dyes act as highlighters for specific targets.

Phase contrast microscopy is used to view biological features in a thin layer of liquid which is usually a buffer solution captured between a glass slide and a thin plastic cover slip.

Visible biological characteristics that can be recognized by microscopy include uniform shape and size. Reflection of light that appears as bright white is another characteristic of biological material.

Fourier transform infrared (FTIR) technology incorporated into microscopes allows for the characterization of chemicals and protein (indicative of biological material) often in powder mixtures. Microscopic use of FTIR is a laboratory technique at this time.

Product Selection

A resource to help you determine which detection products meet your application and need is a document from Edgewood Chemical Biological Center and Joint Program Executive Office for Chemical and Biological Defense entitled *Global CBRN Detector Market Survey*, ECBC-TR-1483, October 2017 by Peter Emanuel and Matthew Caples. The guide assesses commercially available and developing biological detectors as they apply to four areas of application: field use/man portable, mobile laboratory use, diagnostic laboratory use, and high-throughput analytical laboratory use. One hundred seventy-seven biological detectors and/or concepts are rated according to operations, logistics, effectiveness, and biological agents detected. The report also distinguishes which technologies are mature and commercially available versus those not yet commercially available. The information presented in the survey results from a detailed online survey completed by the vendor of each instrument. A web-based companion to the report called *WMD Detector Selector* provides a searchable database version of the extensive 672-page report. Access the web-based version at: https://www.wmddetectorse lector.army.mil/default.aspx.

WMD Detector Selector suggests you determine which of the four areas of application meet your need and then review how the devices or systems were ranked for the application. The scoring and ranking information for each device is presented in U.S. Army-friendly color-coded symbols. Note the raw scores and know that they are somewhat subjective. Next, go to the appendix and read about the specific product in more detail. Finally, contact the manufacturer. The report contains a disclaimer and states that they are not endorsing or recommending any product.

References

Aarino, P., *Expert Systems in Radiation Source Identification*, www.tkk.fi/Units/AES/ projects/radphys/shaman.htm, Helsinki University of Technology, Finland, accessed December 7, 2006,

Advanced Chemical/Biological Integrated Response Course (ACBIRC), *Biological Detection Methods*, Dugway Proving Ground, U.S. Department of Homeland Security, Dugway, UT, August 2007.

ANP Technologies, Inc. website., www.anptinc.com, Newark, DE, accessed December 4, 2006.

Ascher, M., *Development of New Diagnostic Technology Presentation*, Office of Public Health Emergency Preparedness (OPHEP), Washington, DC, August 2002.

Belgrader, P., et al., "PCR Detection of Bacteria in Seven Minutes," *Science* 284: 449–50 (1999).

Belgrader, P., et al., "Rapid Pathogen Detection Using a Microchip PCR Array Instrument," *Clinical Chemistry* 44(10): 2191–2194 (1998).

Bettendorf, A., "Bettendorf Test," *Z. Anal. Chem.* 9: 105 (1870).

Bradford, M.M., A Rapid and Sensitive Method for the Quantitation of Microgram Quantities of Protein Utilizing the Principle of Protein-Dye Binding. *Analytical Biochemistry*, 72: 248 (1976).

Bravata, D., et al., *Bioterrorism Preparedness and Response: Use of Information Technologies and Decision Support Systems*, Evidence Report/Technology Assessment No. 59, AHRQ Publication No. 02-E028, Agency for Healthcare Research and Quality, Rockville, MD, June 2002.

Bravata, D., et al., *Detection and Diagnostic Decision Support Systems for Bioterrorism Response*, www.cdc.gov/ncidod/EID/vol10no1/03-0243.htm, Emerging Infectious Diseases, January 2004.

Brennan, M., *Notes from Conversation*, Owlstone Nanotech, September 29, 2006.

Brockman, S., *Ametek Introduces the Ortec Detective-Ex-100-Latest Advancement in Hand-Held Radiation Identifiers*, press release, Advanced Measurement Technology, Oak Ridge, TN, May 26, 2006.

Brown, C. and Jalenak, W., *From Correspondence*, Ahura Scientific, Chicago, IL, March 14, 2005.

Daum, Keith A. and Fox, Sandra L., *Data for Users of Handheld Ion Mobility Spectrometers*, https://inldigitallibrary.inl.gov/sites/sti/sti/4074867.pdf, U.S. Department of Homeland Security, Idaho National Laboratory, Idaho Falls, ID, May 2008, accessed January 30, 2020.

Drager-Tube/CMS Handbook, 13th ed. Drager Safety AG. KGaA, Lubeck, Germany, 2004.

Eiceman, Gary and Karpas, Zeev, *Ion Mobility Spectrometry*, CRC Press, Boca Raton, FL, 2005.

Engelder, C., Dunkelberger, T., and Schiller, W., *Semi-Micro Qualitative Analysis*, John Wiley & Sons, New York, 1936.

Everett, K. and Graf, F.A., Jr., "Handling Perchloric Acid and Perchlorates," *CRC Handbook of Laboratory Safety*, 2nd ed., Steere, NV, Ed., CRC Press, Boca Raton, FL, January, 1971.

Fatah, A., et al., *Guide for the Selection of Chemical Agent and Toxic Industrial Material Detection Equipment for First Responders,* Second Edition, Guide 100-04, Volume I, National Institute of Standards and Technology, Gaithersburg, MD, March 2005.

Fatah, A., et al., *Guide for the Selection of Chemical Agent and Toxic Industrial Material Detection Equipment for First Responders,* Second Edition, Guide 100-04, Volume II, National Institute of Standards and Technology, Gaithersburg, MD, March 2005.

Feigl, Fritz, *Spot Tests in Inorganic Analysis*, Elsevier Publishing Company, Amsterdam, The Netherlands, 1972. Need ref for ASA ferric nitrate violet.

Fiegl, F., *Laboratory Manual of Spot Tests*, Academic Press, New York, 1943.

Fiegl, F., *Spot Tests in Inorganic Analysis*, 5th ed., Elsevier, New York, 1958.

Field Forensics, Inc. website., fieldforensics.com/index.htm, Largo, FL, accessed January 29, 2020.

Fruchey, I. and Emanuel, P., *Market Survey: Biological Detectors 2005 Edition*, SA-ECBC-2005-01-MSRPT, Edgewood Chemical Biological Center and Critical Reagents Program, Aberdeen, MD, March 2005.

Goldsby, R.A., et al., "Enzyme-Linked Immunosorbent Assay," *Immunology*, 5th ed., W.H. Freeman, New York, 2003.

Goodson, L.H., et al., *Stabilization of Cholinesterase, Detector Kit Using Stabilized Cholinesterase, and Methods of Making and Using the Same*, United States Patent 4,324,858, April 13, 1982.

Gornall, A.G., Bardawill, C.J., and David, M.M., "Determination of Serum Proteins by Means of the Biuret Reaction," *Journal of Biology and Chemistry* 177: 751–766 (1949).

Griffin, R., *Technical Methods of Analysis*, McGraw-Hill, New York, 1921.

Guevremont, Roger and Purves, Randy W., "Atmospheric Pressure Ion Focusing in a High-Field Asymmetric Waveform Ion Mobility Spectrometer," *Review of Scientific Instruments* 70: 1370 (1999).

Guevremont, Roger, et al., "Ion Trapping at Atmospheric Pressure (760 Torr) and Room Temperature with a High-Field Asymmetric Waveform Ion Mobility Spectrometer," *International Journal of Mass Spectrometry* 193: 45–56 (1999).

Guevremont, Roger, et al., *Atmospheric Pressure Trapping of Amino Acids using a FAIMS Ion Trap Coupled to a Quadrupole/Time-of-Flight Mass Spectrometer*, Ionalytics Corporation, Ottawa, ON, Canada, May 2002.

Gutzeit, M., "Gutzeit Test," *Pharm. Ztg.* 24: 263 (1879).

Guyer R.L., and Koshland, D.E., Jr., "The Molecule of the Year," *Science* 246(4937): 1543–46 (1989).

Higuchi, R., Dollinger, G., Walsh, P.S., and Griffith, R., "Simultaneous Amplification and Detection of Specific DNA Sequences," *Biotechnology* 10(4) (1992): 413–7.

Higuchi, R., Fockler, C., Dollinger, G., and Watson, R., "Kinetic PCR Analysis: Real-Time Monitoring of DNA Amplification Reactions," *Biotechnology* 11(9): 1026–30 (1993).

Houghton, Rick, *Field Confirmation Testing for Suspicious Substances*, CRC Press, Boca Raton, FL, 2009.

Innov-X Systems, Inc. website., www.innov-x-sys.com, Woburn, MA, accessed March 14, 2007.

James, R., et al., *Industrial Test Systems, Inc. Cyanide Reagent Strip Test Kit*, ETV Advanced Monitoring Systems Center, Battelle, Columbus, OH, April 2005.

Janeway, C. A., Jr., et al., *Immunobiology*, 5th ed., Garland Science, New York, 2001.

Juhl, W.E., Kirchhoefer, R.D., "Aspirin – A National Survey I: Semiautomated Determination of Aspirin in Bulk and Tablet Formulations and Salicyclic Acid in Tablet Formulations," *Journal of Pharmaceutical Sciences* 69(5): 544–8, May, 1980.

Lachish, U., *Semiconductor Crystal Optimization of Gamma Detection*, Guma Science, Rehovo, Israel, March 1998.

Lowry, O.H., Rosebrough, N.J., Farr, A.L., and Randall, R.J., "Protein Measurement with the Folin Phenol Reagent," *Journal of Biology and Chemistry* 193: 265–275 (1951).

Macherey-Nagel, Inc. website., www.macherey-nagel.com, Easton, PA, accessed January 4, 2007.

Merck KGaA, *Applications for Merckoquant Tests*, photometry.merck.de/servlet/PB /menu/1170820/index.html, Darmstadt, Germany, accessed January 5, 2007.

Michael, Ascher, *Update on Biodetection: Problems and Prospects*, U.S. Department of Health & Human Services, Washington, DC, September 2002.

Mistral Group website., www.mistralgroup.com/default.asp, Bethesda, MD, accessed March 1, 2007.

Mullis, K.B., and Faloona, F.A., "Specific Synthesis of DNA In Vitro Via a Polymerase-Catalyzed Chain Reaction," *Methods in Enzymology* 155: 335–50 (1987).

National Institute for Occupational Safety and Health, *NIOSH Pocket Guide to Chemical Hazards*, Centers for Disease Control and Prevention, Washington, DC, September 2005.

Neubert, H., *Introduction to Matrix-Assisted Laser Desorption/Ionisation Time-Of-Flight Mass Spectrometry (MALDI TOF MS) Department*, King's College London, Strand, England, September 2002. www.kcl.ac.uk/ms-facility/maldi.html.

Ong, Kwok Y., et al., *Domestic Preparedness Program: Evaluation of the TravelIR HCI™ HazMat Chemical Identifier*, Research and Technology Directorate, Soldier and Biological Chemical Command, Aberdeen Proving Ground, Aberdeen, MD, August 2003.

Pasmore, J., *Recent Developments in Handheld X-Ray Fluorescence (XRF) Instrumentation*, Thermo Niton Analyzers, LLC, Billerica, MA, 2009.

Peter, Emanuel and Matthew, Caples, *Global CBRN Detector Market Survey*, ECBC-TR-1483, Edgewood Chemical Biological Center and Joint Program Executive Office for Chemical and Biological Defense, Edgewood, MD, October 2017.

Pibida, L., et al., *Results of Test and Evaluation of Commercially Available Radionuclide Identifiers for the Department of Homeland Security, Version 1.3*, National Institute of Standards and Technology, Gaithersburg, MD, May 2005.

Pibida, L., Karam, L., and Unterweger, M., *Results of Test and Evaluation of Commercially Available Survey Meters for the Department of Homeland Security, Version 1.3*, National Institute of Standards and Technology, Gaithersburg, MD, May 2005.

Pibida, L., Karam, L., and Unterweger, M., *Results of Test and Evaluation of Commercially Available Personal Alarming Radiation Detectors and Pagers for the Department of Homeland Security, Version 1.3*, National Institute of Standards and Technology, Gaithersburg, MD, May 2005.

Pier, G.B., Lyczak, J.B., and Wetzler, L.M., *Immunology, Infection and Immunity*, ASM Press, Washington, DC, 2004.

Precision Labs UK Company website., www.precisionlabs.co.uk, King's Lynn, United Kingdom, accessed January 4, 2007.

Raman, C., and Krishnan, K., "A New Type of Secondary Radiation," *Nature* 121: 501 (1928).

Reinking, L.N., Reinking, J.L., and Miller K.G., "Fermentation, Respiration and Enzyme Specificity: A Simple Device and Key Experiments with Yeast," *American Biology Teacher* 56: 164–168 (1994).

Roberson, R., *Correspondence and Notes from Conversation*, Sensidyne, Clearwater, FL, February 2007.

Saiki, R.K., Bugawan, T.L., Horn, G.T., Mullis, K.B., and Erlich, H.A., "Analysis of Enzymatically Amplified Beta-Globin and HLA-DQ Alpha DNA with Allele-Specific Oligonucleotide Probes," *Nature* 324(6093): 163–66 (1986).

Sandia National Laboratories, *Miniature Ion Mobility Spectrometer*, www.sandia.gov/mstc/technologies/microsensors/IMS.html, accessed November 16, 2006.

Sensidyne, Inc., *Sensidyne Gas Detector Tube Handbook*, Clearwater, FL, undated.

Shriner, R., and Fuson, R., *Systematic Identification of Organic Compounds – A Laboratory Manual*, John Wiley & Sons, New York, 1935.

Smith, E., and Dent, G., *Modern Raman Spectroscopy – A Practical Approach*, John Wiley & Sons, West Sussex, England, May 2006.

Smith, P.K., et al., "Measurement of Protein Using Bicinchoninic Acid," *Analytical Biological Chemistry* 150: 76–85 (1985).

Sun, Yin and Ong, Kwok Y., *Detection Technologies for Chemical Warfare Agents and Toxic Vapors*, CRC Press, Boca Raton, FL, 2005.

U.S. Department of Energy, National Nuclear Security Administration, *Evaluation of Preventative Radiological/Nuclear Detector Archetypes to Validate Repurpose to the Consequence Management Mission*, U.S. DOE Brookhaven National Laboratory, 2017, https://www.osti.gov/servlets/purl/1425190, accessed February 7, 2020.

U.S. Department of Homeland Security, *Radiation Dosimeters for Response and Recovery Market Survey Report*, Science and Technology Directorate, National Urban Security Technology Laboratory, 2016, https://www.dhs.gov/science-and-technology/saver, accessed February 2, 2020

U.S. Department of Justice, *An Introduction to Biological Agent Detection Equipment for Emergency First Responders, NIJ Guide 101–00*, Office of Justice Programs, National Institute of Justice, Washington, DC, December 2001.

U.S. Department of Justice, *Color Test Reagents/Kits for Preliminary Identification of Drugs of Abuse*, NIJ Standard-0604.01, National Institute of Justice, National Law Enforcement and Corrections Technology Center, Rockville, MD, July 2000.

United States Army, *Chemical Agent Detector Kit, M256A1*, www.army.mil/fact_files_site/m256a1/index.html, U.S. Army Fact File, Washington, DC, accessed February 2, 2007.

United States Army, *Operator's Manual for Chemical Agent Detector Kit M256/ M256A1*, U.S. Army Technical Manual 3-6665-307-10, September 1985.

United States Army, *WMD Detector Selector*, Edgewood Chemical and Biological Center, https://www.wmddetectorselector.army.mil/default.aspx, accessed February 6, 2020.

Wayne, C., *A Simple Qualitative Detection Test for Perchlorate Contamination in Hoods*, www.orcbs.msu.edu/chemical/resources_links/contamhoods.htm, Office of Radiation, Chemical and Biological Safety, Michigan State University, East Lansing, MI, accessed January 6, 2006.

Yalow, R.S., and Berson, S.A., "Immunoassay of Endogenous Plasma Insulin in Man," *Journal of Clinical Investigation* 39(7): 1157–75 (1960).

Strategies 4

Figure 4.1 Choose a strategy that allows you to safely characterize the hazards of an unknown material.

Overview

This chapter provides strategies for identifying hazards of unknown materials. When conducting a qualitative analysis, consider circumstantial clues for individual cases and integrate the information into your analysis. Your overall strategy should use as many resources as possible. Confidence in a result will be strong when information is verified from three sources, although this may be challenging to accomplish (Figure 4.1).

Additional instrumentation is becoming available for field qualitative analysis. If you have specialized equipment available to you, always confirm a result produced by one particular piece of equipment. A strategy for characterizing the hazards of unknown material should include confirmation from at least one, preferably two, additional forms of technology. Remember, a competent technician is the most important part of the emergency characterization of unknown materials.

While explosive material may be characterized in this analysis, collection of material from an intact explosive device is extremely dangerous and should be referred to an explosives expert.

If you focus completely on invoking and observing reactions of the unknown substance in a quest to find answers, you risk your

The most efficient test scheme for an unknown material is to field test it with Raman, FTIR, and HPMS instrumentation. Once identified, confirm the result with at least one orthogonal technology.

If more in-depth testing is necessary or if no identification results from Raman, FTIR, and/or HPMS, refer to Chapter 5 for information on manipulating samples that are not readily identified by an identifier instrument.

If no result is produced through sample manipulations refer to Chapter 4, which provides additional strategies for characterizing and identifying hazards of unknown materials.

Approximately 400 specific field tests are available in Field Confirmation Testing for Suspicious Substances, CRC Press.

safety and the safety of others in the area (Figure 4.2). Operating safely at all times is paramount to emergency characterization of unknown substances. Your safety is more valuable than any benefit that might be realized by disregarding safe analysis methods in order to save time or effort. A simple set of rules will help you work safely (Figure 4.3).

Commandments of Thy Characterization Strategy

- Thou shalt be careful when handling all chemicals, especially if they be toxic or explosive.
- Thou shalt acquaint thyself with all manner of safety devices and clothe thyself in protective raiment.
- Thou shalt cause neither fire nor explosion apart from thy analysis, nor be guilty of any other kind of accident.
- Thou shalt recall as long as thou shalt toil, that surely as thy mother brought thee into the world, she, as well as the hazards of thy labor, can take thee out.

Figure 4.2 "I think we need a better plan." Use of a sound strategy for emergency characterization of unknown materials is essential for safe operation.

- Thou shalt not suffer thy hand to confuse hot objects as cool.
- Thou shalt abide an open mind, and ye shall suffer to communicate when ye become vexed or befuddled.
- Thou shalt seek help by calling out, or by activating thy personal alert safety system, or by walloping thy panic button.
- Thou shalt not feast nor quaff in thy workspace.
- Thou shalt not work alone, no matter what thy rank.
- Thou shalt evacuate thy workspace when warning bellows from above or below or beside thee.
- Thou shalt protect all manner of people, and shalt not suffer little children to come into thy hot zone.
- Thou shalt report all chemical misdeeds, even those that do not currently besiege thee, to thy boss.
- Thou shalt control thy supply of reagents and all manner of expensive detection equipment, and thou shalt dispose of thy chemical wastes according to the laws of thy land.

Figure 4.3 Heed ye now the good words of the Commandments of Thy Characterization Strategy.

Insofar as ye shall observe all these commandments, ye shall continue to receive the blessings of thy higher authority commensurate with thy pay. Insofar as ye shall break this covenant, then surely shall thy workspace be as ashes and ye will have characterized thy last sample sooner than forsoothed.

Choosing a Strategy

Every analysis of unknown material is unique based on the sample, the situation in which it is presented, the resources available to process the sample, and the skill of the technician. The majority of cases involve common industrial hazardous material and situations in which there is no criminal intent to cause harm. Situations that are intended to cause injury, death, or fear happen much less frequently but tend to present much higher risk to the analyzing technician and those exposed to the unknown material. In either case, some ancillary information may be available that can help you select a strategy. This chapter presents several strategies that can manage samples from the "basic" unknown material as well as agents used as weapons of mass destruction, explosives, and illegal drugs.

The strategies employ a logic that helps you characterize the material by defining its hazards (e.g., explosive, corrosive) or grouping it with other similar substances (e.g., nerve agents). In some cases, the substance might be identified by name, in which case you can access additional resources for information and continue testing to confirm the result and increase confidence in the identification. Whether the substance is characterized or identified, the purpose of this book is to help you understand the hazards presented by the unknown material so that you can take appropriate actions to protect life, property, and the environment during an emergency or in the workplace. You should become familiar with all the strategies before choosing one to begin your analysis. As new information becomes known, you should be able to change strategies as necessary. Always begin with initial observations, and then choose a characterization strategy based on your observations, circumstantial evidence, and any other clues. The basic characterization strategies are for analysis of suspected:

- Unknown Material
- Chemical Warfare Agent
- Biological Warfare Agent
- Radiological Agent
- Explosive Material
- Illegal Drug

Initial Observations

To begin the analysis, take a moment to look at the material and make some notes. Remember to record all findings as you proceed through the analysis, including any identifying characteristics of the container.

Always eliminate immediate threats from a closed container:

- **Gamma radiation survey.** Use a radiation monitor to detect gamma radiation. Since radiation testing is noninvasive and can detect an otherwise undetectable hazard, a radiation screen should always be done before sample manipulation begins if there is any possibility the unknown sample could be radioactive. If gamma radiation is detected at less than twice background, proceed with the next paragraph. Otherwise, proceed with analysis as described below in Characterization of Suspected Radiological Material.
- **Crystallized or gelled material.** If the container is transparent, look for crystal or gel formation with a light source on the opposite side. Examine the container opening for crystal formation; if crystals are found, consider them potentially explosive. If you decide to continue analysis, gently test the crystals for peroxide. To test for

peroxide crystals on or near the lid, gently rinse the material with a solution of 50% acetone in water without disturbing the container (Figure 4.4). Collect this small amount of liquid and test for oxidizers. You may be able to wash the material directly onto a peroxide test strip. If the peroxide test strip is negative, test the material around the lid with a pH test strip, a potassium iodide starch strip, a nitrate test strip, and methylene blue solution. Continue testing based on your findings.

- **Increased temperature.** Determine the temperature of the container with an infrared thermometer. A container that is warmer than ambient conditions is under pressure. Do not use an infrared thermometer with a sighting laser if the sighting laser can penetrate the container wall and potentially explode the contents. This is unlikely if using a class-2 laser sight typical of an IR thermometer that generates less than 1 mW of power. In contrast, a Raman spectrometer may use a class-3B laser with an output of up to 450 mW of power that can ignite combustibles such as darkly colored paper, charcoal, or gunpowder. Refer to the user manual.

Determine a safe method of opening the container, if necessary. You are responsible for the safe manipulation of the substance. Be sure that the act of opening the container does not shear crystals formed in the lid threads or otherwise cause a reaction. You may need to open a container remotely, in an inert gas or liquid, by piercing the sidewall or some other method dependent on your unique situation.

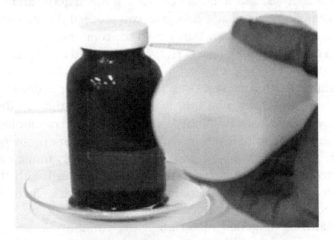

Figure 4.4 Peroxide material dried between the cap and container can shear and explode when the lid is removed. A small volume of acetone and water gently applied under the container lid will dissolve exposed solid material. The rinse solution can then be tested for peroxide.

Unknown Material

This strategy for characterizing hazards of unknown material is the most general strategy and will be used most commonly. It contains broad screening tests while the other strategies contain tests focused on specific hazards. While performing the broad screening tests in this strategy, be aware that you may need to adopt a new strategy based on your findings. For example, if you ignite a sample in this strategy and it burns intensely, like a road flare, you should consider elements of the strategy for characterizing explosive material.

Characterizations in Air

Characterizations in air are determined from tests done in the open air. Materials that are liquids or solids are removed from their containers and observed. Gases can be released into a sample bag for easier handling. They can also be bubbled through water and tested at the water surface; however, do not test gas directly from the cylinder to prevent reaction in the tubing or container. If testing a gas, see the section that follows. If testing a liquid or solid, continue below.

Solids and Liquids

Open the container if you determine it to be safe. Observe the contents for a moment to determine if any reaction with air is occurring. Watch for and note vapor production, spontaneous ignition, color change, or any other indication of a physical or chemical reaction.

If the material is a liquid, estimate vapor pressure by using water (vapor pressure about 20 mmHg) as a standard and the classifications below. You may find it helpful to place a drop of water next to a drop of liquid sample on an inert surface, set it aside for a few minutes, and then observe the evaporation rate.

- Low (oil-like): <10 mmHg
- Water-like: ~20 mmHg
- Moderate: 30–100 mmHg
- High: >100 mmHg
- Gas: >760 mmHg

Test any vapor above the material with a wet pH test strip, then a wet potassium iodide starch strip, and then a nitrate test strip (use a nitrate test strip only if the vapor pH is acidic). Shake excess water from the strips so water does not drop into the container. Do not make contact with the contents of the container; the material will be contacted directly with a smaller, safe amount in subsequent testing.

If the sample is a solid or liquid and not suspected of being explosive, place a gram of material (about the size of an M&M candy) on an inert surface and record any reaction. (First conduct the hammer and anvil test and flame test described in the *Characterizing Explosive Material* section if the material may be an explosive.) A heavy glass ashtray works well for this test because it is inert, resistant to shattering, easy to wipe clean, and inexpensive. Of course, labware such as a watch glass may be used. While any reaction is still occurring, test for gas or vapor just above the material with a pH strip and a potassium iodide starch strip. If no reaction occurs above the sample, move the test strips directly onto the sample. If the sample reacts violently with the water on the test strips, document your findings and continue carefully since many subsequent tests involve water.

Test for flammable gas or vapor emitting from the material by attempting to ignite the gas or vapor with a wand lighter (Figure 4.5), then move the flame directly onto the sample and try to ignite the liquid or solid. Use caution. A gram, about the size of an M&M candy, is a practical amount of sample to use in this hazard characterization scheme; however, a gram can be a dangerous amount of material if it is an explosive. Explosives can produce several thousand calories of heat per gram.

You might note some characteristics in the way the sample reacts to the flame, but this is a safety step that screens explosive and flammable material. More detailed combustion characteristics will be observed later.

Gases
Consider a radiation screen if you suspect the gas could be radioactive. Radioactive gases are not likely to be encountered, but a gamma radiation screen can be performed on a small sample in a bag from outside the bag. A

Figure 4.5 Ignition of combustible vapors from a water-reactive substance using a wand-style lighter.

small stream of gas flowing from the bag can be screened for alpha and beta radiation. If the gas is also suspected of being air reactive, you may need to perform the alpha and beta screen in an inert atmosphere, such as a glove box filled with nitrogen. Consider the control of potentially radioactive gas that has been released from the cylinder.

Open the container if you determine it to be safe. Observe the contents for a moment to determine if any reaction with air is occurring. Watch for and note spontaneous ignition, color change, or any other indication of a physical or chemical reaction. If the gas begins burning on exposure to air, refer to the section on air-reactive hydrides in Chapter 2.

Test the gas with a wet pH test strip, then a wet potassium iodide starch strip, and a nitrate test strip if the pH is acidic.

Sample the gas with a combustible gas indicator in the presence of adequate air to determine flammability. An alternative test for combustibility is to bubble the gas through water and use a wand lighter to ignite the bubbles as they leave the water. Figure 4.6 depicts a schematic view of a device that can be used. Do not attempt to ignite gas in or from a sample bag or gas emitting directly from a cylinder.

Evaluate the characteristics of the gas and continue with a gas qualitative analysis system (e.g., Sensidyne Deluxe Haz Mat III) if more information is needed. Gas testing concludes at this point.

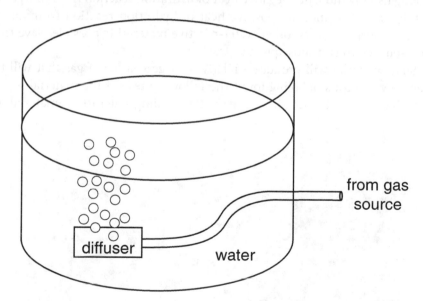

Figure 4.6 Bubble sample gas through water when testing for flammability. The water will prevent flame from spreading into the cylinder. Use a diffuser at the end of the tubing in the water to produce small bubbles. If the gas is observed to be acidic, be careful that the water does not inadvertently boil when gas is being bubbled through the water.

Characterizations in Water

Many reactions can be observed by adding the sample to water in one of two methods. If the material may react violently with water, place a gram of material on an inert surface and add water from the side. Use this open method if a material produces a gas upon contact with water. For less reactive solids, add about ½-inch or 1 cm of water to a test tube and add a gram of solid sample to the water. Note the specific gravity, any dissolving, color change, etc. and then agitate with a pipette and observe again. If the sample is liquid, mark the surface of the water on the test tube before adding an equal volume of liquid by slowly running it down the side of the test tube. Be careful to point the tube in a safe direction as contents of the tube may eject if a violent reaction occurs. After you are certain a violent reaction will not occur, gently agitate the contents with a pipette and avoid injecting air into the liquid.

Determine the solubility of the liquid and whether it is heavier or lighter than water. Estimate solubility of a liquid sample by determining the new location of a visible division between liquids (Figure 4.7). No division practically indicates miscibility. You may need to mix the sample with the water by gently squeezing a pipette at the base of the tube. Do not shake the tube; shaking can cause splashing or cause air to be mixed into the sample. Measure the temperature of the liquid to determine change in temperature (Figure 4.8). Cooling usually indicates negative heat of hydration reaction from inorganic salts. Heating can indicate positive heat of hydration reaction from acids, bases, or inorganic salts; other water-reactive material is possible. Save this water solution to run subsequent tests.

Some material will produce a relatively large volume of gas that will be evident by vigorous bubbling (omit the following tests if the material is not producing a gas). Characterize this hazard by adding water to the material on

Figure 4.7 Marking the water level on the side of a test tube (left) and then mixing gently with a pipette (center) will help you estimate solubility (right), which is about 30% in the case of this lighter-than-water liquid.

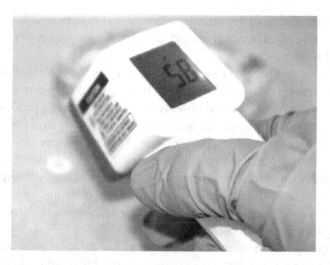

Figure 4.8 Use an infrared thermometer to measure changes in temperature relative to ambient conditions.

an open, inert surface and then holding a lit match just above the material. Flammable gases will burn intermittently as they are produced. Hydrogen gas will pop with an invisible flame. Acetylene will flash with a deep orange flame and produce black soot in the air. Carbon dioxide and other asphyxiants will extinguish the flame. Corrosive gases can be detected with a pH strip wetted with deionized water and held above the substance. Toxic hydrogen sulfide and hydrogen cyanide can be detected with colorimetric air monitoring tubes. Oxidizing gases can be detected by holding a wetted potassium iodide starch test strip above the wet substance. Some of these gases can be detected with electrochemical sensors in various models of air monitors. Test a dilute sample when using an air monitoring instrument and check the cross sensitivity of the relevant sensor to prevent false identification of the gas. Some colorimetric air monitor tubes are designed to screen for these gases. You can use Table 3.12, Sensidyne Deluxe Haz Mat III Detection of WRTIH Gases, in the previous chapter, as one method of analysis for toxic inhalation hazards from water-reactive material as defined by DOT. Non-TIH gases such as carbon dioxide can be identified by individual gas tubes as suggested by the water reaction.

pH is determined by combining the unknown substance, water, and a pH test strip. If the unknown is a solid, the pH strip should be dipped into an aqueous solution made from the solid. A small amount of liquid may be tested directly from a pipette; carefully add the unknown to a small amount of water if you are not sure if water is present. If the unknown is a gas, wet the pH strip with deionized or distilled water, place it in the gas, and then allow time for the gas to dissolve into the water and affect the pH dyes. This method for gas is not entirely accurate near neutral pH because the water tends to fix the pH dyes immediately; however, it is dependable at extremely acidic and basic pH. Do not reuse pH test strips. Save this water solution for subsequent tests.

Some acids and bases are more volatile and pose a greater inhalation risk. To estimate the volatility of the sample, hold a wetted pH test strip over the sample on an open surface and observe the time it takes to produce a result. Some substances, such as glacial acetic acid, react slowly and may take up to two minutes to fully produce colors on the pH strip; however, most will react quickly (e.g. ammonium hydroxide) and provide an estimate of volatility.

If the pH strip colors bleach to pale or white, it is likely the unknown is a strong oxidizer. If the pH strip is destroyed by "melting" or if it blackens, the unknown is a powerful, reactive substance or is strongly dehydrating and most likely indicates concentrated sulfuric acid.

A screening test that detects all oxidizers is a difficult prospect. As mentioned in Chapter 3, testing by oxidizing potential is relative. Also, oxidizers react in differing methods and colorimetric tests do not detect all methods. The following screen for oxidizers will indicate the presence of most oxidizers.

1. Use a potassium iodide starch test strip to detect most oxidizers. This test covers a broad variety of oxidizers, namely anything that can oxidize iodide from potassium iodide to free iodine. The free iodine in the presence of starch in the test paper will turn blue to black. Very powerful oxidizers will produce a quick flash of blue color followed immediately by bleaching to white; watch closely for this result as it happens very quickly and should be interpreted as a positive result.

2. If the previous step is negative, test the sample with a peroxide test strip. This will detect most, but not all, organic peroxides.

3. If the previous step is negative, test the sample with a nitrate test strip. Nitrates and nitrites will not react with the previous steps. This test will detect nitric acid and nitrate anion found in substances such as ammonium nitrate.

4. If the previous step is negative, test the sample with a 0.3% aqueous solution of methylene blue. If a purple color results from the original deep blue, perchlorate is present above 0.001 M.

5. Oxidizers that do not produce a positive result in the previous four steps will need to be determined with substance-specific tests or in other tests such as burning. Negative results in the previous four steps strongly suggest an oxidizer is not present.

If an oxidizer is found to be present, consider that the unknown may be an explosive material. You can reference the oxidizers in explosives by consulting a table in the strategy for explosive material (later in this chapter) before further testing.

The types of compounds determined in this oxidizer screening test describe a large number of individual substances. However, there are more common compounds that are more likely to be encountered. There are three groups of common materials listed below that can polymerize, form peroxides, or do both. The most hazardous compounds are the first group, those that form peroxides on storage without being concentrated. These are materials that can accumulate hazardous levels of peroxides simply in storage after exposure to air. Substances in this first group should always be checked for peroxides. The obvious test, a peroxide test strip, is an excellent indication of hazardous peroxide formation. Unfortunately, the peroxide test strip will not work for all peroxides or substances that only present a polymerization hazard. Therefore, a negative result on the peroxide test strip does not preclude the possibility of peroxide or polymerization hazard and other testing is necessary. The screen for peroxide is described below.

1. Carefully examine the container for date and overall condition. If possible, contact manufacturer for guidance.
2. Containers that show signs of oxidation (rust, bleached colors, etc.), contain crystals or gelled material, or are stored longer than the recommended shelf life should be handled with extreme caution. Do not open or move any of these containers until you have consulted an expert.
3. Quantitative test strips (e.g., EM Quant Peroxide test strip from Merck) can be used to determine the presence of some peroxides. Another method is the oxidation of iodine from potassium iodide through the use of a potassium iodide starch strip. A more sensitive technique uses 100 mg of potassium iodide in 1 ml of glacial acetic acid added to 1 ml of the unknown liquid in a test tube or graduated cylinder. The appearance of yellow or brown color indicates peroxide >0.01%.

In this first group, the usual peroxide test will only work for isopropyl ether. A concentration of less than 80 ppm is considered safe. If any of these substances are suspected, even though a peroxide test strip is negative, other characteristics can be observed, such as crystal formation or gelling. Consult an MSDS or manufacturer before opening a container if one of these compounds is suspected. If a sample has already been obtained, further tests suggested below may be performed on small amounts. These materials will present observable characteristics in contact with water that identify hazards or suggest further simple testing:

- Sodium amide: Reacts violently with water to produce ammonia and sodium hydroxide (vapor and solution will be strongly basic).

- Potassium amide: Similar to sodium amide.
- Vinylidene chloride: Insoluble and heavier than water liquid that is volatile and flammable. Vapors will be positive with a Beilstein test or extremely diluted with a refrigerant detector (use a very diluted air sample) and combustion products will be acidic (pH strip). To perform the Beilstein test, place a small copper wire in a torch flame until it does not discolor the flame. Introduce sample vapors to the air inlet of the torch. Iodine, bromine or, most likely, chlorine are indicated if the flame turns green or blue. The Beilstein test yields false-positive with acids and some nitrogen compounds (e.g., urea).
- Isopropyl ether: Moderately volatile, water-insoluble, flammable liquid similar in appearance to acetone (peroxide test strip will detect peroxides formed by isopropyl ether).
- Potassium metal: Will slowly form a yellow-orange crust of peroxide (shock sensitive) upon exposure to air. Potassium metal will react violently with water to leave basic potassium hydroxide solution (pH strip) and will produce hydrogen gas that will self-ignite and burn with an invisible flame; however, potassium contaminant in the hydrogen flame can produce a light purple flame.
- Butadiene: Stored in a cylinder as a flammable liquefied gas that can form spontaneously combustible peroxides if exposed to air (this gas may burn spontaneously if exposed to air or the container may explode spontaneously if air is introduced to the pressurized container).
- Divinyl ether: Hazardous characteristics similar to butadiene.
- Divinyl acetylene: Hazardous characteristics similar to butadiene.

This second group of solvent materials can form hydroperoxides and ketone peroxides that can be a hazard if heated or concentrated. These peroxides are soluble, and detection is straightforward with peroxide test strips or potassium iodide starch test strips.

- Furan
- Isopropanol
- Methylcyclohexane
- Methyl isobutyl ketone
- Tetrahydrofuran
- Tetrahydronaphthalene
- Dicyclopentadiene
- Diglyme
- Diethyl ether (ethyl ether)
- 1,4-Dioxane
- Ethylene glycol dimethyl
- Ether (monoglyme)
- Acetal
- 2-Butanol
- Cellosolves (e.g., 2-ethoxyethanol)
- Cumene (isopropylbenzene)
- Cyclohexene
- Decalin (decahydronaphthalene)

This third group of materials includes liquids and liquefied gases that can be initiated by oxygen to polymerize. A peroxide test strip may not detect all peroxides that might be present, but if the liquid shows an increased viscosity or if the gas leaves a residue, suspect the presence of a polymer or peroxide. These materials are safe to use or store if the peroxide concentration is less than 80 ppm.

- Acrolein: A colorless to yellow, volatile flammable liquid that is 40% soluble in water that will readily polymerize without an inhibitor. To determine solubility, add a few milliliters of water to a test tube or graduated cylinder, mark the top surface and then add an equal volume of the unknown liquid. Mix thoroughly, but be aware that the contents may eject from the container. Allow the mixture to settle and determine the solubility based on any line of separation that forms between the two liquids.
- Acrylic acid: Flammable, water miscible organic acid (use pH strip) that freezes at 55°F and readily polymerizes upon loss of its inhibitor. Acrylonitrile: Lighter than water, slightly soluble, flammable liquid. Hydrogen cyanide gas can be detected with a colorimetric air monitoring tube after a small amount of liquid is ignited. Chloroprene (2-chlorobutadiene): chlorinated (refrigerant detector or Beilstein test) flammable liquid but can polymerize if not inhibited. Chlorotrifluoroethylene: also known as R1113, a chlorinated (refrigerant detector or Beilstein test), flammable (acidic smoke determined by pH strip) liquefied gas that can polymerize.
- Ethyl acrylate: Appears as a common flammable liquid but can polymerize without an inhibitor.
- Ethyl vinyl ether: Appears as a common flammable liquid but can form peroxides on storage (use peroxide test strip).
- Styrene: A clear to yellow oily liquid, lighter than water and insoluble that can polymerize if heated or upon the loss of its inhibitor. As an aromatic, a small sample will burn with an orange flame and produce black soot in the flame.
- Tetrafluoroethylene: A flammable liquefied gas, also known as R1114, used to make Teflon® that can polymerize upon loss of inhibitor. The Beilstein test does not work on fluorocarbons and refrigerant detectors are less sensitive to fluorocarbons. A colorimetric air monitoring tube can detect hydrogen fluoride in the combustion products, which will be acidic (use pH strip).
- Vinyl acetate: A flammable liquid that can polymerize upon loss of its inhibitor.

Characterization in Flame

When initially testing a liquid, soak a cotton swab with the unknown liquid. Hold the swab horizontally and bring a match or lighter flame toward it from a line slightly lower than the swab. Relative flammability can be estimated by the distance fire flashes between the swab and flame. Combustible liquids will require heating by the lighter flame before it ignites. Either flammable or combustible liquids will burn on the cotton swab and sustain combustion when the lighter is removed. The liquid will burn away and the cotton should remain unburned for some time. Water-based liquids will obviously not light and the water will quench the cotton even when the lighter flame is held under it (Figure 4.9). Mixtures of water and organics will ignite relative to concentration. Some of these mixtures may flash as the organic portion is distilled away from the water.

You can test a larger sample of liquid if needed. Place a small sample, 1 to 2 inches in diameter, on a flat, nonreactive surface. Ignite the sample with a lighter flame moved toward it from the side.

If you are interested in fire suppression actions, you may want to place a sample of the material aside after covering with Class B firefighting foam, typically a 3% mixture, and set a timer so you can estimate when a foam blanket will need to be refreshed. A 3% foam mixture can be made by adding one drop of concentrate to 33 drops of water and agitating in a test tube.

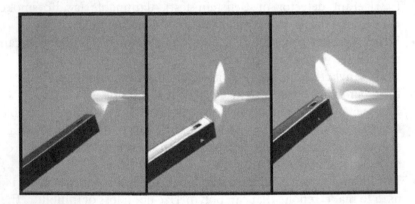

Figure 4.9 Characterization of a fire hazard from a liquid sample on a cotton swab. Left: A water-based liquid that does not contain a high concentration of polar organic solvent (e.g., acetone, methanol) will not burn. Note that the cotton swab is not burning because the heat of the flame is quenched by the water. Middle: A combustible liquid with a high flash point (charcoal lighter in the image) must be heated directly by the flame before the sample sustains combustion. Right: A combustible liquid with a low flash point (petroleum ether in the image) will flash as the flame appears to jump through the air.

Vigorous agitation will inject air into the foam. This method is especially helpful when a material cannot be named for reference in the foam manual. Use of perfluoroalkyl and polyfluoroalkyl substances (PFAS) has been restricted, but not all Class B foams contain PFAS.

To screen flammable, combustible, explosive, and other highly thermally energetic materials, which may be liquids or solids, place a very small amount of material on a flat, inert surface and hold the flame on it. As described in the *Characterizing Explosives* section, start with a milligram and increase sample size if an explosive is suspected. Never exceed 1 gram. A wand-style lighter works well for this purpose. A propane torch flame will work, but it can also blow away the sample. Increase the size of the sample when you are confident a larger sample will not explode. Larger samples can be burned on a wire or spatula in a torch flame. Observe flame and smoke characteristics similar to those listed in the flammable liquids section. Sparks may indicate a metal (Figure 4.10). Test smoke for pH and oxidizers. When using test strips, colorimetric air monitoring tubes or other detection devices, be sure to sample smoke that has cooled to a manageable temperature to avoid damage to the detector.

Solid samples can be burned by selecting a method most appropriate to the material. A small sample can be heated directly by the flame of a lighter over an inert surface. Alternatively, a sample can be held on a spatula, wire loop, or other noncombustible device and held in a low-velocity propane torch flame or ignited with a lighter flame. Energetic material may explode or burn intensely; noting the rate of combustion will help you determine hazards. You can also burn solid material by heating a gram of sample in a test tube with a propane torch. This method will cause intense smoke production and the smoke can be ignited by quickly moving the torch flame into the smoke and then back to the base of the test tube. Never use a test tube if explosive material is suspected to prevent glass fragmentation.

Materials that burn intensely, pop, or otherwise react energetically in the flame should be suspected as explosive material, especially if they are solids. Oxidizing smoke from these materials is also reason to suspect an explosive. Some explosive material will burn vigorously but not explode unless slightly confined or compressed. Never confine a burning material while characterizing flame reactions.

Test the smoke for pH and oxidizers with wetted test strips. Test for toxic gases if suspected. When using test strips, colorimetric air monitoring tubes or other detection devices, be sure to sample smoke that has cooled to avoid damage to the detector. Table 4.1 lists several results and indications of burning material and Table 4.2 describes elemental flame colors. Table 4.3 summarizes characteristics of explosive material revealed in testing.

Figure 4.10 Metal dust forming sparks in a propane torch flame.

Characterization of Toxins

Everything is toxic dependent on dose. Characterization of toxic material is very difficult because there is obviously no field test for toxicity. The easy answer is that everything is toxic and you should avoid all contact with unknown material; however, that is not a practical response during an emergency. If an unknown material can be identified, toxicity data from references can be used to formulate a plan of action. Identifying an unknown material to predict its toxicity would obviously be the strategy of choice, but it is often not achievable.

This section describes simple and quick field tests that can identify potential toxic hazards from unknown material. These tests will emphasize the production of gases and vapors from unknown materials that pose a serious inhalation hazard.

Colorimetric air monitoring tubes designed as a screening matrix are very useful for determining vapors and gases emitting from an unknown material. These gases could be produced spontaneously or may be produced

Table 4.1 Flame and Smoke Characteristics

Observation	Characteristic
Clear to black or sooty smoke	Simple alkane structures burn completely and produce little smoke. More black (carbon) smoke is produced by complex structures, aromatics, and increasing number of carbon-carbon double and triple bonds. Transitioning smoke characteristics indicate a mixture.
Orange flame with blue base	Simple organic material.
Orange flame; black, sooty smoke	Aromatic compounds.
Orange-yellow flame with blue-green portion; black smoke	Chlorinated hydrocarbons.
Pink or purple in flame	Amine, nitrile, or organic nitrate.
Purple	Iodine or organo-iodine compound.
White on or in flame	Organo-sulfur or organo-phosphorous compound.
Yellow flame, possibly with bright white base; white, sooty smoke	Silane-based substance.
Yellow or blue flame color partly transparent and difficult to see in sunlight; soft flame edges	Alcohols.
Yellow to red smoke	Yellow can indicate iodine or organic nitrates.
Yellow, orange, or blue flame with various colors	A flammable liquid that could be contaminated by inorganic compounds. Refer to Table 4.2.

Table 4.2 Elemental Color in Propane Flame

Element	Flame Color
Arsenic	Blue
Barium	Yellow-green
Boron	Green
Calcium	Orange-red
Copper	Green
Lead	Gray-blue
Lithium	Red
Potassium	Light purple
Sodium	Yellow-orange
Magnesium dust	White sparks
Aluminum dust	White-yellow sparks
Zinc dust	Blue sparks
Iron dust	Red-orange sparks

upon exposure of the unknown material to another material. In any case, these tubes are very helpful in characterizing or identifying unknown gases and vapors.

Multiple sensor monitors are generally not useful in identifying gases and vapors because they are often presumptive and prone to interferences.

Table 4.3   Characteristics of Explosive Material Revealed by Qualitative Tests

Analysis Step	Characteristic*
Ignition as described for flammable solids	Hydrocarbon
Ignition as described for flammable solids	Nitro group
Ignition as described for flammable solids	Organic nitrate
Nitrate test strip	Nitrate anion
Methylene blue	Perchlorate
Peroxide test strip, potassium iodide starch test strip	Peroxide
Heavier than water	Metal
Oxygen sensor in a multisensor detector	Oxygen
Temperature	LOX (liquid oxygen)
Ignition as described for flammable solids	Picrate
Ignition as described for flammable solids	Unstable structure

*Refer to Table 4.5 as necessary.

An exception is the combustible gas sensor that will characterize a gas or vapor as flammable.

Some aqueous solutions can produce toxic gases due to a shift to acidic pH. Paper test strips are useful in identifying these solutions. Sulfide test strips can detect sulfide compounds that will generate hydrogen sulfide gas if the pH is forced lower. Cyanide test strips will detect soluble cyanide compounds that will generate hydrogen cyanide gas if the pH is forced lower. Hypochlorite solutions (e.g., bleach) will produce chlorine gas if the pH is forced lower; however there is no reliable hypochlorite test strip. Hypochlorite solution should be expected if it turns a potassium iodide starch strip black and then bleaches it white. Various colorimetric test strips described in Chapter 3 are available for the detection of specific toxins and are susceptible to test limitations and interferences. Table 3.10, Sensidyne Colorimetric Test Sets for Qualitative Analysis, also summarizes tests for unknown gases.

Characterization of Certain RCRA Wastes

Some workers might find certain tests helpful for assigning Resource Conservation and Recovery Act (RCRA) hazardous waste codes. Spot tests are described in Chapter 3 for the following:

- Arsenic
- Barium
- Cadmium
- Chromium

- Lead
- Mercury
- Selenium
- Silver

It is important to note that these tests are not approved by the EPA due to insufficient sensitivity. These tests are useful for bulk sampling; if a test is positive, the element is present in the sample and must be considered when assigning hazardous waste codes. If a spot test is not sufficient for your application, refer to other technologies that might be applicable.

Generally, Raman spectroscopy is effective at detecting complex ions containing metals, but is not effective at detecting elements and single atom cations. For example, Raman would likely detect potassium chromate, but is unlikely to detect chromium chloride. Infrared spectroscopy can detect many individual compounds, but contaminants, especially in the form of waste, make a library match difficult.

X-ray fluorescence (XRF) has strong application in the detection of metals. XRF is used in some EPA methods. Consult the instrument manual to determine if detection limits are low enough for your application.

Characterization of Halo-Organics

Test for chlorinated organic material using a thin copper wire or a thicker wire hammered flat that has been heated until orange in a propane torch flame. While holding the red hot copper in the torch flame, hold the sample near the air intake holes for the torch head. You might need to gently heat the sample in order to volatize it. The sample vapors will enter the torch head and will be ionized in the flame. Chlorine will combine with the copper to form volatile copper chloride, which will produce a characteristic blue or green torch flame.

Chlorinated hydrocarbons are the most common of the halo-organics, and many are harmful to the environment due to their persistency. This test will also detect organic bromine and iodine, but not fluorine (copper fluoride is not volatile). Interference is caused by some nitrogen-containing material such as urea, pyridine, and oxyquinolines, which will form volatile copper cyanide and produce a green flame color. The presence of some acids, especially halogen acids, produces a green flame color.

Organic iodine will produce purple smoke when burned; yellow smoke in the presence of water vapor. Organic bromine produces a nonoxidizing yellow smoke. Bromine and iodine materials tend to be very dense.

Fluorides can be detected with a spot test. Organic fluorine must be converted to inorganic fluoride, usually by reaction with a strong base. The fluoride spot test described in Chapter 3 that utilizes alizarin can detect fluoride at low ppm concentration.

Identification of Unknown Material

Emergency characterization of unknown material will help you identify hazards. The hazards should then be confirmed by additional testing to increase confidence. Comparatively, a result produced by a modern detector should be confirmed by another test method. The presence of three independent sources indicating the same result produces the highest confidence expected in field testing.

Various detection technologies may be useful in confirming hazards associated with the characterization tests described previously. You may be able to use a characterization test to confirm a result produced by a detector. For example, if a combustible gas indicator alarms, a sample of a liquid found nearby can be tested in a remote place to determine if the combustible gas indicator (CGI) alarm was produced by the liquid instead of some other source. Or a liquid found to be combustible in the characterization test should be tested with a CGI to determine if the monitor reacts to the particular substance.

Chemical Warfare Agents

Characterization of chemical warfare agents are separated into four classes:

- Nerve agents
  - G-agents
  - V-agents
  - A-agents
- Blister agents
  - Sulfur mustards
  - Nitrogen mustards
  - Lewisite
- Choking agents
  - Phosgene and diphosgene
- Blood agents
  - Hydrogen cyanide
  - Cyanogen chloride
  - Arsine

Intentional release of these materials will most likely be characterized by signs and symptoms of those exposed. If this is the case, the following characterization methods can be used to confirm the signs and symptoms. If exposure has not occurred, characterization methods will be necessary before further actions are taken.

Consider the characterization strategy in three scenarios: one that has significant signs and symptoms present in humans and other susceptible organisms, one that is completely isolated from human contact, and finally one that involves an accumulation of the precursor chemicals and equipment nessessary to produce chemical warfare agents clandestinely.

There are many field-portable devices designed to warn the operator of the possible presence of a chemical warfare agent. These detectors are designed to alarm at a very low concentration of what is assumed to be a chemical warfare agent. While these instruments have a valid place in first response and domestic preparedness, they are more helpful in characterizing a material as a possible chemical warfare agent rather than actually identifying the agent. A cache of unfamiliar chemicals may only reveal the potential presence or production of a chemical warfare agent once the analysis clues are connected and cross referenced. As you consider your strategy, compare available instruments as to their function: identifying vs. detecting and warning.

Characterization of Nerve Agents

Characterization of an unknown substance as a nerve agent is primarily dependent on its ability to inhibit cholinesterase. Many improvised compounds can be used as weapons of terror, and your ability to determine cholinesterase allows you to take immediate actions. Additional testing is needed to characterize the sample as a G-agent, a V-agent, and now, an A-agent (novichok).

In the case of human exposure to a nerve agent, the typical SLUDGE symptoms (salivation, lacrimation, urination, diaphoresis, gastrointestinal distress, and emesis) will be the primary signs for characterization. Miosis (pinpoint pupils) is also an early sign. Although some symptoms may be delayed up to 18 hours, the more likely presentation is a few minutes through skin contact and almost immediately through inhalation exposure. In this scenario, the signs and symptoms are enough to characterize the substance as a nerve agent.

Confirmation tests for cholinesterase from Chapter 3 may be used to screen cholinesterase inhibiting materials. The Agri-Screen Ticket (5 minutes) can be used for solid, liquid, and wipe samples and the M256A1 kit can be used for vapor samples (20 to 25 minutes). These tests are very sensitive and specific to cholinesterase inhibiting materials, including nerve agents and many commercial pesticides.

Flame photometry technology is the choice of many organizations that need a detection method for nerve agents. Detectors such as the AP2C detect phosphorous and sulfur compounds in air; however, this method does not differentiate G-, V- or A-series nerve agents from some commercial pesticides, specifically the thiophosphates (malathion, diazinon, chlorpyrifos, etc.) and does not detect carbamates.

Colorimetric air monitoring tubes are screening tests for nerve agents. Tubes that detect "nerve agents" are tests for phosphoric acid esters, not specifically for nerve agents. As such, this test does not detect nonphosphoric acid ester materials that inhibit cholinesterase, many of which make up the majority of commercial pesticides.

In the case of a suspected nerve agent that has not caused symptoms, the confirming technology results become the main indicator of nerve agents. In this case, confirm inhibition of cholinesterase with a colorimetric test.

To determine if a chemical warfare nerve agent is present, use Table 1.32, Elemental Content of Nerve Agents to select tests that would differentiate nerve agents. Hydrolysis of G-agents will produce characteristic substances that can help differentiate G-agents. Hydrogen chloride, hydrogen fluoride, and hydrogen cyanide may be produced by hydrolysis. Detection of fluorine in a cholinesterase inhibitor is strongly suggestive of a G-agent because commercial pesticides do not contain fluorine. Likewise, nitrile (cyanide) is strongly suggestive of nerve agents. Hydrogen chloride is unlikely from commercial pesticides because most chlorinated pesticides have been banned due to their persistency in the environment.

Hydrolysis of V-agents will produce strongly acidic solutions and decomposition products that are nearly as powerful cholinesterase inhibitors as the original agent. Interestingly, VX is 3% water soluble at 77°F but miscible below 49°F. This trait is detectible in a sample of pure VX, but the addition of solvents, thickeners, or contaminants will affect the solubility.

The M256A1 kit will detect these nerve agents as vapor, but the low vapor pressure of the sample may make vapor detection in a hood or other ventilated area difficult. Placing the test in a small enclosure with the sample would concentrate vapors.

In the 1970s the Soviet Union began research and development of a new class of nerve agent, now known as A-series nerve agents. These new nerve agents were developed under the codename novichok or "newcomer." Novichoks were developed into three distinct analogs; A-230, A-232 (withstands cold temperatures), and A-234 (a solid derivative). By 1990, the Soviet Red Army had adopted agent A-232 for use in all munitions designated for CW agent delivery.

The physiological effects of novichoks are similar to other nerve agents. However, the acetylcholinesterase inhibition associated with novichoks is considered irreversible. Pharmacological treatment is also consistent with other nerve agent exposure consisting of anticholinergics, anticonvulsants, and acetylcholinesterase reactivating agents.

Chemical characteristics of novichoks are similar to G-agents sarin and soman. Novichoks are considered to be five to eight times more toxic than VX. The very low vapor pressure of the A-series agents makes them persistent and difficult to detect.

Identification of Nerve Agents

Other confirming or identifying results can be obtained from more advanced technologies. Gas chromatography mass spectrometry excels at identifying chemical weapons and their precursors and/or degradation products because all chemical weapons are volatile organic compounds; especially if the chemical warfare agent is in a solvent or reaction mixture, for example, sarin in acetonitrile in the Tokyo subway incident. Field-portable instruments include the backpack sized Inficon HAPSITE GCMS and the hand-held MX908® HPMS.

IMS has been the traditional form of detection for these agents over the last twenty years. IMS detectors include the LCD 3.3®, Chempro 100i IMS ®, RAID-M 100, IMS-2000®, GID-3, and Sabre 5000®. FAIMS is now available in the Owlstone NGCD.

Different versions of SAW detectors are being developed and tested. Improved selectivity and miniaturization are improving performance and in combination with a gas chromatography (GC) column may make these instruments a good choice for CW agent detection.

Flame photometry can indicate the possibility of a G-agent if phosphorous alone is detected, or a V-agent if both phosphorous and sulfur are detected. A-agents would alarm as G-agents on conventional instruments before U.S. declassification of A-agent information. Flame photometry instruments such as the AP2C and MINI-CAMS® are stable, sensitive field detectors, which makes them likely to be considered during emergency characterization of unknown material. However, without the use of a GC column, these detectors cannot provide positive identification and can only operate in a generic manner. Detection of phosphorous could indicate a commercial organophosphate (DDVP, methamidophos, mevinphos). Detection of sulfur and phosphorous could indicate commercial thiophosphates (aspon, chlorpyrifos, diazinon, malathion, methyl parathion, phosmet).

Infrared spectroscopy instruments as well as photoacoustic spectroscopy and Fourier transform infrared spectroscopy instruments can detect bulk amounts of CW agent, but lack the sensitivity to detect vapor and trace amounts. A bulk sample of nerve agent will most likely contain production by-products and sample concentration could be a problem because a solvent and contaminants from wipe samples would complicate spectral interpretation. FTIR instruments include the TruDefender®, Gemini, and HazmatID 360® among others. Raman spectroscopy can detect CW agents with some of the same limitations, but with the ability to sample through an intact clear container. Raman instruments include FirstDefender®, Gemini, and Progeny ResQ® among others. The Thermo Gemini® was able to reliably detect VX, GA, GB, and GD in bulk testing, but accuracy dropped when samples contained less than 12% agent. The Gemini and First Defender instruments will detect these agents at lower concentrations using an optional analysis technique (called tagging) that causes the instrument to interrogate the spectrum for specific chemical compounds, however it will not detect trace amounts.

Trace detection in the field is possible with portable GC/MS instruments such as the Inficon HAPSITE and HPMS instruments such as the MX908. These instruments will detect volatile organic compounds at the parts per million and parts per billion sensitivities.

A-series nerve agents have recently been declassified in the United States. Characterization and identification of A-series agents are similar to G- and V-series agents. A-series agents were developed by the Soviets in the 1970s in part to sidestep Chemical Warfare Convention treaties established by the Organization for the Prohibition of Chemical Weapons (OPCW) to reduce the proliferation of chemical weapons. The first documented use of A-series agents occurred in 2018 at a Salisbury, England park. Positive identification was made through trace analysis of residual A-series agent.

The physiological effects of novichoks are similar to other nerve agents. However, the acetylcholinesterase inhibition associated with novichoks is considered irreversible. Pharmacological treatment is also consistent with other nerve agent exposure consisting of anticholinergics, anticonvulsants, and acetylcholinesterase reactivating agents.

Chemical characteristics of novichoks are similar to G-agents sarin and soman, but include an acetoamydin radical $(C_2H_5)_2N-C(CH_3)=N-$ bound to the phosphorus molecule instead of the O-alkyl radical. Novichoks are considered to be five to eight times more toxic than VX. The very low vapor pressure of the A-series agents makes them persistent and difficult to detect. Signifiantly, the composition of A-232 and A-234 circumvented the provisions of the Chemical Warfare Convention (CWC). Novichoks were added to the list of chemicals banned by the CWC in November 2019. These compounds are known or referred to as novichoks, fourth-generation agents (FGA), nontraditional agents (NTA), and A-series agents. Specific chemical information vital to identifying novichoks has recently been declassified by the United States and concurrently with addition to the CWC list.

Field-portable IMS instrumentation would not likely be a good choice for A-series agent identification due to the lack of vapor production.

A field sample of an A-series agent for testing by FTIR or Raman indicates production or storage; dispersed agent will not provide enough material for a Raman or FTIR test. Raman and FTIR instruments are capable of identifying bulk amounts of A-series agents if the instrument manufacturer has procured and distributed library spectra for each agent that consider states of production, storage, dispersal, and degradation.

HPMS is capable of identifying A-series agents if collected with a swab and processed through a thermal desorption unit and the analysis algorithm has been programmed to recognize characteristic ions resulting from the instrument's ionization source. A range of agents may be recognized, similar to recognition of several fentanyl derivatives.

GC/MS is capable of identifying nanogram amounts of A-series agents; however, extremely low volatility is a challenge to the operator. Applying the appropriate method, column, and even carrier gas configuration will optimize the analysis. Comparing A-series agent GC/MS chromatograms to published spectral libraries (i.e., NIST and AMDIS) will not yield a direct match until declassified information is added to the libraries. In this case, the operator could search a chromatogram for the most abundant or unique ion fragments associated with A-series agents.

Interestingly, M8 paper delivers a characteristic color change unique to A-series agents. A drop of A-series agent placed on M8 paper will either fully or partially absorb into the paper and result in a yellow/gold color change (indicating a G-agent). In about ten seconds, the yellow/gold color will transition to green/blue color (indicating a V-agent). The color transition may be caused by the A-series similarity to both G- and V-agents with an aging reaction within the M8 paper.

Characterization of Blister Agents

Blister agents described here are all heavier than water liquids that are barely soluble in water. They include sulfur mustard, nitrogen mustards, and Lewisite. The physical description of the material and the colorimetric detection methods are generally not sensitive enough to detect these materials below IDLH levels.

Sulfur mustards are characterized as insoluble, heavier than water liquids that are clear if pure but are more likely amber to black. Colorimetric air monitor tubes are available for the detection of thioether. Specificity is increased if chlorine is also detected. Flame photometry (AP2C) can detect sulfur if the vapors can be drawn into the detector; sample heating through the use of an accessory or some other method may be necessary.

Nitrogen mustards are characterized as "sparingly" soluble, heavier than water liquids that appear as clear to yellow. Nitrogen mustards may also be solids. There are amines and colorimetric tubes available for amine detection; however, some amine tubes are simply pH indicators showing the presence of a basic gas. A true amine detection tube would greatly increase specificity.

The organic chlorine might be detected with a refrigerant detector, but not at very low levels that can still be toxic.

Lewisite is characterized as an insoluble, heavier than water liquid that may appear as clear to brown. It contains arsenic; a Drager colorimetric tube can detect organic arsenic compounds and arsine. Lewisite can break down and release acetylene. A Sensidyne acetylene tube (101S) can detect acetylene, but the reaction occurs due to any substance that can reduce molybdate. An XRF detector could be used to identify arsenic in a sample, but would not identify the material as Lewisite.

Blister agents may be less obvious in the short term if human exposure has occurred. Lewisite and sulfur mustard produce pain immediately. Nitrogen mustard pain may be delayed a few hours.

The M256A1 kit will detect these blister agents as vapor, but the low vapor pressure of the sample may make vapor detection in a hood or other ventilated area difficult. Placing the test in a small enclosure with the sample would concentrate vapors. The organic chlorine might be detected with a refrigerant detector, but not at very low levels that can still be toxic.

Other confirming or identifying results can be obtained from more advanced technologies as individual circumstance requires. GC would be helpful in separating mixtures; especially if the CW agent is in a solvent or reaction mixture.

Identification of Blister Agents

IMS has been the traditional form of detection for blister agents over the last 20 years. IMS detectors include the LCD 3.3, Chempro 100i IMS, RAID-M 100, IMS-2000, GID-3, and Sabre 5000. FAIMS is now available in the Owlstone NGCD.

Different versions of SAW detectors have been developed and tested. Most notably is the Joint Chemical Agent Detector (JCAD). Improved selectivity and miniaturization are improving performance, and combination with a GC column may make these instruments a good choice for CW agent detection as they undergo evaluation.

Different versions of SAW detectors are being developed and tested. Improved selectivity and miniaturization are improving performance, and combination with a GC column may make these instruments a good choice for CW agent detection as they undergo evaluation.

Flame photometry can indicate the possibility of a G-agent if phosphorous alone is detected or a V-agent if both phosphorous and sulfur are detected. Flame photometry instruments such as the AP2C and MINI-CAMS are stable, sensitive field detectors, which makes them likely to be considered during emergency characterization of unknown material. However, without the use of a GC column, these detectors cannot provide positive identification and can only operate in a generic manner.

Infrared spectroscopy instruments as well as photoacoustic spectroscopy and Fourier transform infrared spectroscopy instruments can detect bulk amounts of CW agent, but lack the sensitivity to detect vapor and trace amounts. A bulk sample of blister agent will most likely contain production by-products and sample concentration could be low since a solvent and contaminants from wipe samples would complicate spectral interpretation. FTIR instruments include the TruDefender, Gemini, and HazmatID 360. Raman spectroscopy could also detect CW agents with some of the limitations already discussed, but with the ability to sample through an intact glass container.

Trace detection of blister agents in the field is possible with portable GC/MS instruments such as the Inficon HAPSITE and HPMS instruments such as the MX908. These instruments will detect volatile organic compounds at parts per million (PPM) or parts per billion (PPB) levels.

Characterization of Choking Agents

Choking agents include phosgene and diphosgene. Phosgene is a gas and diphosgene is a liquid that slowly decomposes to phosgene during extended storage. Phosgene can be detected with several colorimetric air monitor tubes. Detection of diphosgene is not as common. Both gases will be detected by a refrigerant detector. Both gases can decompose and form hydrogen chloride gas that can be detected with a pH strip. The M256A1 kit does not detect either gas/vapor.

Identification of Choking Agents

Choking agents will be analyzed in air, and there has not been as great an effort to develop instruments to identify these agents, in part due to the greater effectiveness of other chemical warfare agents. Beyond the characterizing tests listed above, laboratory analysis of an air sample is most likely necessary due to low demand and the lack of field equipment capable of such identification.

Other confirming or identifying results can be obtained from more advanced technologies. GC would be helpful in separating mixtures, especially if the CW agent is in a solvent or reaction mixture. Various forms of advanced technology are available for the detection of phosgene, but data are lacking concerning diphosgene. Check the SAVER database to determine if an instrument that might be available for characterizing an unknown substance is applicable.

Characterization of Blood Agents

Characterizing blood agents requires air testing technologies. Blood agents include three gases: hydrogen cyanide, cyanogen chloride, and arsine.

Hydrogen cyanide is lighter than air and soluble in water; more so at high pH. Hydrogen cyanide can be detected in air by several colorimetric air monitor tubes and the M256A1 kit. Hydrogen cyanide in water can be detected by colorimetric test strips after conversion to cyanogen chloride.

Cyanogen chloride is a gas above 55°F but may be a liquid or solid in cold temperatures. It has a high vapor pressure and is heavier than water and 7% soluble. Cyanogen chloride may be detected directly as a gas, vapor, liquid, or solid using a wet cyanide test strip that uses the cyanogen chloride detection method. The test strip method is not sensitive enough to detect cyanogen chloride in air concentrations near IDLH, but it may be acceptable for hazard characterization. Several colorimetric air monitoring tubes are available, some using the same chemistry as the test strips. The M256A1 kit does not detect cyanogen chloride in air.

Arsine is a heavier than air, arsenic-based gas that may be detected by several colorimetric air monitor tubes. Several tests are available for use in the tubes and several are cross sensitive to phosphine. It is unlikely that an XRF instrument could detect arsenic in arsine in a dilute atmospheric sample. If a sample could be concentrated, XRF might discriminate arsine from phosphine.

Identification of Blood Agents

Blood agents will be analyzed in air and there has not been as great an effort to develop instruments to identify these agents, in part due to the greater effectiveness of other chemical warfare agents. Beyond the characterizing tests listed above, laboratory analysis of an air sample is most likely necessary due to low demand and the lack of field equipment capable of such identification. Identification is likely to be presumptive initially.

Other confirming or identifying results can be obtained from more advanced technologies. GC would be helpful in separating mixtures; especially if the CW agent is in a solvent or reaction mixture. Various forms of advanced technology including photoionization detectors (PIDs) are available for the detection of these gases. Check the SAVER database to determine if an instrument that might be available for characterizing an unknown substance is applicable.

Biological Warfare Agents

Field screens for biological material are used in an environment where some assumptions can be made. For example, a volume of "white powder" must be large enough and concentrated enough to be seen. An amount the size of a pin head would be adequate to test for protein at concentration of 1% or so with a reliable protein test. This amount is well above the volume of biological material that can be lethal, but it is a practical consideration of field characterization.

Characterization of biological warfare agents includes screening tests and field identification tests for biological material. Although field identification tests, by name, are intended to presumptively identify an agent, CDC states laboratory culture is the only acceptable method of completely reliable identification of biological warfare agents.

Characterization of Biological Warfare Agents

Screening for biological material includes the greatest number of possibilities, but is the least specific. It should be used as an immediate yes or no answer to the presence of biological material. Field identification testing should be used as a yes or no answer to the presence of a specific biological warfare agent or a group of agents, depending on the test. Culture is used to grow and identify specific agents based on observation and laboratory testing over a few days. Screening is the most inclusive and least accurate at identifying agents. Culture is the most accurate but takes days. Presumptive field identification testing is specific to certain agents, not totally dependable, and takes 15 to 30 minutes to complete the test.

Characterization of a suspected biological warfare (BW) agent should begin with a review of the situation and threats present. First responders and others have protocols in place for management of "white powder" incidents that should be followed. For example, if a threat letter mentioned anthrax and the letter accompanied a powder, it would be a good idea to use a field identification test for anthrax. In most cases, there is no specific threat. Use the following characterization scheme for material that is suspected to be biological material and adapt it as influenced by the threat analysis of an individual incident.

In the case of a powder or liquid, take a few moments to observe the material. Brilliant white or bleached powders are not likely to be biological agents; they are more likely to be inorganic material such as silica or a bleached food product such as enriched flour or corn starch. Biological agents are most likely to be tan to dark brown to almost black, depending on purity and how they were processed. Brightly colored material may be a marker that indicates the presence of a fungicide or other agricultural product enhancement.

The more finely a material is divided (Figure 4.11), the greater respiratory hazard it presents. Solids that are less than 5 microns (a human air is about 75 to 100 microns thick) have buoyancy properties that allow them to drift in air and to be inhaled deeply into the lungs. Particles this small will also develop

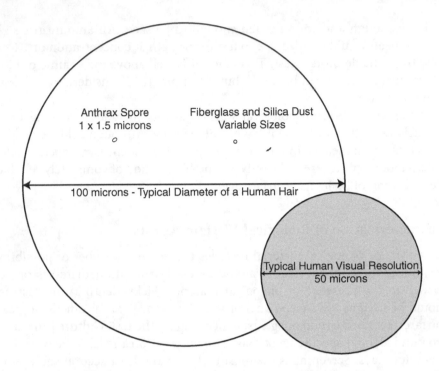

Figure 4.11 Biological agents are more effectively dispersed in air when the particle size is small.

electrostatic charge that will cause them to clump together, which decreases their effectiveness as a weapon. Particles that can be seen jumping from static charge when disturbed have not been engineered to drift freely in air.

Remember to split the sample to preserve a reasonable amount of material for evidence and further testing by other organizations. Test an amount about the size of a pinhead for protein using a protein test. The protein test utilizing bicinchoninic acid (BCA) is most efficient. If the test is negative, the material contains less than 1% protein and is highly unlikely to contain a biological agent. If you suspect less than 1% BW agent mixed into an inert material (which would greatly degrade its effectiveness as a weapon), continue with specific identification testing and submit for culture based on threat analysis. If the protein test is positive, the material contains protein and might be a BW agent, but is statistically likely to be a food product. Material that produces a positive result in the BCA protein test is either a protein material (food or BW agent) or a material that can reduce copper in the test.

Perform a starch test. If the starch test is positive, the material contains starch and is likely to be a food product. If you suspect a BW agent in starch (which would greatly reduce its effectiveness as a weapon), continue with specific identification testing and submit for culture based on threat analysis. If the starch test is negative, no starch is present. The material is now also known to contain protein material (food or BW agent) or a material that can reduce copper in the test. Likely food materials include protein substances such as yeast, protein powder, malt powder, etc. Presence of protein and lack of starch increase suspicion of a biological agent.

Additional observations include the following clues. When added to water, particles that dissolve easily are much less likely to be a biological hazard than those that float on the water or move up the wall of the test tube. A small amount of biological material, if touched with a red hot piece of metal, like a paperclip, should char and smoke similar to flour or sawdust.

Biological Warfare Agent Production Awareness

Situational awareness will help characterize biological warfare agent production before attempting to identify the agent. A biological warfare agent production facility may appear as a large combination of components, but the function of the components is somewhat similar for production of most agents, varying mostly by the quality of the final product. The following list of signatures may be useful in recognizing a clandestine biological lab and its intended function.

- Petri dishes and broth tubes indicate bacteriologic or possibly myco-logical (fungal) production.
- Growth on a petri dish is likely bacterial but could also be yeast or mold.
- Most bacterial colonies are white, cream or yellow in color and approximately circular in shape (Figure 4.12).
- Yeast colonies, a type of fungi, appear similar to bacterial colonies but some grow as white or yellowish patches with a glossy surface (Figure 4.13).
- Molds are fungi and they often appear whitish-grey with fuzzy edges. They usually turn into a different color, from the center out-wards (Figure 4.14).
- Slant tubes with broth or chicken eggs with small openings in the shell may indicate viral production.
- An improvised fermenter may appear as a container of foul-looking liquid.
- Canning jars or other sealed containers under pressure may indicate anaerobic growth of *Clostridium botulinum* for production of botu-lism toxin. Putrid, rotting meat in the container is a telltale sign.
- D-glucose (dextrose) or other sugars can be used as a carbon source.
- Antifoam is used to prevent protein from foaming in a fermenter.

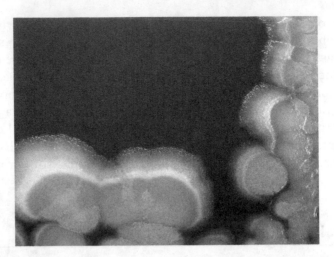

Figure 4.12 10x magnification of colonial growth of Sterne strain members of the Gram-positive bacterium, *Bacillus anthracis* cultured on sheep blood. (Todd Parker, PhD, Associate Director for Laboratory Science, Division of Preparedness and Emerging Infections at CDC.)

- A ceramic bead mill in a rock tumbler can be used to mill some toxins, some virus, and especially spores such as those of *Bacillus anthracis*.
- Any vacuum device can be a dissemination device if the electrical polarity is reversed.
- A glove box or hood may be large enough to contain a setup for the full process of fermentation, concentration, drying, and grinding of a biological agent. Key components include:
  - An improvised fermenter.
  - A funnel with a filter on an aspiration (side arm) flask connected to a vacuum. The aspiration set up serves to remove most of the liquid from the growth in the broth by vacuuming the broth through a filter, thus removing most of the liquid and collecting solid material on the filter.
  - A toaster oven can be used to dry the paste collected in the filter paper.
  - A cardboard box or other similar container with an incandescent light bulb or other mild heat source could work as an incubator.
  - A coffee bean mill or a mortar and pestle can be used to coarsely grind the dried agent.
  - A rock tumbler with a ceramic bead mill inside can be used as the final milling step. Addition of Cabacil® or Aerosil®, types of light and fluffy silica, can occur in this step.

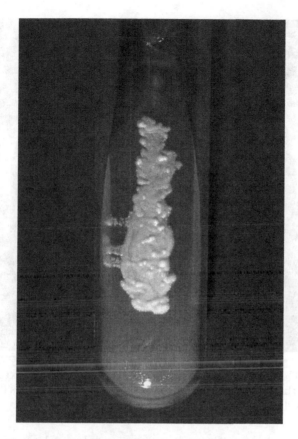

Figure 4.13 A single slant culture test tube with the yeast-form of the fungal organism *Sporotrichum schenckii*. The colony is characteristically smooth but knobby and slightly yellow. (CDC/ Dr. Kaplan.)

- A flask with growth separate from fermenters may contain a starter culture for the fermenters. A haziness, cloudiness, or turbidity of the broth can indicate growth.
- A fairly clear solution may contain dissolved toxins. A hazy, cloudy or turbid liquid is a suspension that may indicate bacterial or fungal cells suspended in liquid.
- Gram stains are used to microscopically view vegetative cells in the fermenter. Reagents used to dye Gram-positive and Gram-negative bacteria include crystal violet (methyl violet), iodine, Safranin O (basic red 2) and Fuchsine (fuchsin, rosanilin, rosaniline hydrochloride).
- Malachite Green is a stain used to microscopically view spores.
- Ammonium sulfate or polyethylene glycol (PEG) is used to purify proteins by salting in or salting out of the solution. Some proteins fall out of the solution as a precipitate while others may stay in the liquid

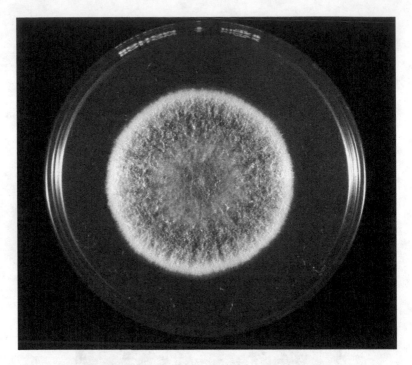

Figure 4.14 The fungus *Aspergillus flavus*. Note the roughly textured, granular appearance of the colony, and becoming darker in its coloration as it began to age, from the central region outward. (CDC/ Dr. Lucille K. Georg.)

(the supernatant). Other salts or compounds can be used, but these are the most likely.

- T-flasks are used for animal cell culture, which is necessary to support the growth of virus and rickettsia.
- Most biological materials require a dissemination device or vector for delivery to a target population. Improvised dissemination devices include refillable aerosol containers, spray bottles, Class A fire extinguishers, hollow ceramic balls, etc. Vectors may include rodents, mosquitos, fleas, etc.
- Laboratory personal protective equipment or decontamination supplies are clues to hazards as perceived by the operator.

Biological agents from a clandestine production site could contain material other than the agent if the material is in production or if purification methods are lacking. Fermentation mixtures may include egg whites (albumin), blood, sugars, etc. Fermentation mixtures are likely to be liquids similar to any common fermentation process, such as beer brewing. Freeze drying might be used to preserve the agent, regardless of its purity.

Identification of Biological Warfare Agents

If the material is found to contain protein and not starch, the likelihood of a biological agent is increased. The next level of field testing would utilize specific immunoassay tests for specific agents or groups of agents. Since this characterization scheme assumes an amount of sample that can be seen, be careful not to overload the identification test; consult instructions for specific tests. There is no method that would reduce the number of individual tests that would need to be performed based on the tests themselves. If a threat or some other clue accompanies the material, use an individual test based on the clue. Identification tests simply give a positive or negative result for the organism, group of organisms or toxin that the test is designed to identify. Field tests are susceptible to false-positive results; continue testing all the possible identities and retest any positives to increase confidence. Performing serial dilutions is a method for reducing the incidence of false-positives. Serial dilutions require the tests to be run in triplicate, one each for the dilutions of 1:1, 1:10, and 1:100. A true presumptive positive for immunoassays occurs when all three dilutions indicate positive results.

Polymerase chain reaction (PCR, also known as nucleic acid amplification) could also be used at this point in the scheme for its high sensitivity and specificity. PCR testing takes slightly longer to run (30–60 minutes) but yields much more specific and reliable results based on DNA amplification. Disadvantages of PCR include the need for pathogen-specific probes, which may not be available. Additionally, a specific probe must be used for each pathogen; a PCR screening test would be problematic. PCR may not work on toxins because depending on the synthesis they may not have a nucleic acid sequence, but toxins contain protein and would be detected by the BCA protein test and by specific immunoassays.

Regardless of the results produced by the tests above, current CDC policy requires samples of material suspected to contain biological warfare agents be submitted to the laboratory resource network (LRN) for further testing. Culture is the only method capable of identifying a biological warfare agent with certainty. Some materials submitted as a possible biological material will not respond to culture if they are not an organism. For example, ricin is a toxin that contains protein, but will not replicate since it is only a large chemical. LRN is expanding capability to identify toxins and chemicals that are not biological material by working with chemical analysis counterparts.

Radiological Material

Characterization of radiological material is accomplished only through the proper use of radiological detection instruments. Unknown materials that present significant radiation hazard are characterized quickly so that appropriate actions can be taken. This characterization method uses a threshold of twice normal background radiation at the location of the test. If you are attempting to detect lesser or trace amounts of radiation for the purpose of identifying radionuclides for evidence or needs other than emergency categorization of unknown material, consult a radiological expert.

Characterization of Radiological Material

If an unknown substance is not inside a container, it may be screened for alpha, beta, and gamma radiation with a wide spectrum monitor operated in the appropriate modes or by using a combination of monitors that span the spectrum. Rather than search for a quantifiable measurement, a relative response of twice background or more is a reasonable threshold for radiation hazard screening. Determine the threshold by doubling normal background radiation for the geographic area in which the test is occurring (Figure 4.15 and Figure 4.16). This makes the test relative and allows the technician some judgment in determining what is hazardous.

A relative response monitor that detects alpha, beta, and gamma radiation would be able to detect a source equal to background radiation by displaying a value roughly twice that of background. If no significant increase is detected, the sample is not a significant radiation hazard. A significant

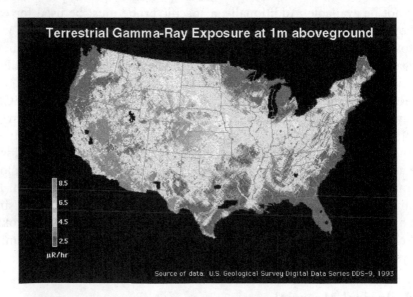

Figure 4.15 Normal geographic background radiation for the continental United States as determined by the U.S. Geological Survey (www.usgs.gov). Determine normal background radiation of your location with the radiation instrument used in the analysis.

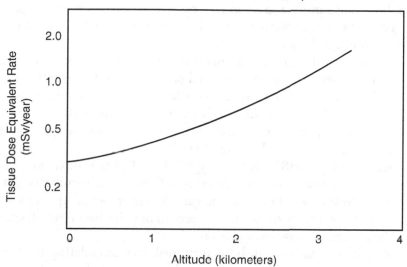

Figure 4.16 Human cosmic radiation exposure increases with altitude due to lower shielding effect from the atmosphere. Changes in altitude can affect the value you use for normal background radiation. (Adapted from NCRP Report 94, Exposure of the Population in the United States and Canada from Natural Background Radiation.)

gamma hazard would be detected while approaching the source. Alpha and beta hazards would be detected as the sensor is moved close to the source.

Containerized samples may be screened for gamma radiation through the container. An exception might be a gamma source contained in a vessel designed to prevent most of the radiation from leaving the container. If this is the case, the container will be constructed of very dense, thick sidewalls. Recognition of a container designed to safely store a powerful source is a major clue regarding the presence of a radiation hazard.

To characterize the sample, configure a gamma radiation monitor in a remote place. The technician must understand the monitor output as it relates to background radiation. Approach the container and place the sensor very close to the sidewall of the container. If the container is partially filled, place the sensor near the area of the container that holds the unknown material, not the headspace. If the background count does not increase, there is no significant gamma radiation being emitted through the container.

Closed containers should only be opened if other hazards have been reasonably eliminated, such as explosive or shock-sensitive material. Containerized samples of alpha and beta emitters will most often need to be opened in order to expose the sensor to potential radiation. When the container is opened, the gamma screen should be repeated.

To begin the radiation hazard screen, configure a monitor that detects alpha, beta, and gamma radiation in a remote place. This is often the same monitor used to screen the container for gamma radiation. Determine normal background radiation.

The monitor will count radiation events and the technician must be able to estimate an average reading. If normal background is ten counts per minute (CPM), an analog meter using a needle display may bounce above and below 10 CPM. CPM may be displayed as mRem/hr. The audio output, usually a click or beep, will not sound exactly every six seconds (60 seconds divided by 10 CPM), but will average 10 CPM.

Place a gram (an M&M candy weighs about 1 g) of sample on an inert surface. Using the most common detector, a Geiger-Mueller tube, orient the monitor so the tube window faces the sample one or two centimeters above it. If the monitor detects radiation by more than twice background, characterize the material as a radiation hazard.

The type of radiation can be determined by manipulating the sensor. Most radiation sources will be gamma or high energy beta, which can penetrate the sidewall of the tube. Orient the side of the GM tube one-quarter to one-half inch above the unknown material. If the count remains two or more times background, suspect a gamma or high energy beta source. If the count falls when the side of the GM tube is oriented just above the sample, suspect a low-energy beta or alpha source.

Reorient the GM tube window about a half-inch above the sample. The count should return to the abnormally high reading originally found. Slide a piece of paper between the source and the GM tube window. If the count decreases, the unknown material likely contains an alpha source.

Reorient the GM tube window about a half-inch above the sample. The count should return to the abnormally high reading originally found. Slide a piece of one-eighth inch thick aluminum between the source and the GM tube window. If the count decreases, the unknown material likely contains a beta source.

If the unknown substance contains, as its only radioactive source, a pure low-energy beta emitter, the radiation hazard may not be detected by a GM tube. Examples of low-energy beta emitters are hydrogen-3 (tritium), carbon-14, and sulfur-35. These specific sources are difficult if not impossible to detect in a field screen using only a GM tube. These low-energy beta emitters are detectable only by more specialized and sensitive detectors, such as liquid scintillation counting instruments or open window gas proportional detectors. If a radioactive hazard is suspected due to incident circumstance, these low-energy beta hazards should not be neglected. Such equipment may be available from another agency on a case by case basis if only a low-energy beta emitter is suspected. Curie quantities of tritium (a relatively large amount) do not present an external exposure hazard because the low-energy beta emissions cannot penetrate the outer layer of skin. There remains an inhalation and ingestion hazard because tritium is distributed throughout the body as water.

Most common isotopes will emit a combination of alpha, beta, and gamma radiation and can be detected with a common GM tube. The most likely combination is a mixture of beta and gamma radiation. By using this method of radiation screen, serious gamma sources will be found as the technician approaches the closed sample container. Weaker gamma sources will be found as soon as the GM tube is placed near a sample. Serious beta and alpha sources will be found as soon as the GM tube window is oriented above the sample.

If a radiation detector is capable of sensing alpha, beta, and gamma radiation and utilizes both visual and audible outputs, the technician is immediately alerted to increases above background. Serious beta and alpha sources are discovered quickly and the technician has the option of leaving the area or, in the case of smaller increases, of differentiating between alpha and beta sources.

When determining the ratio of alpha, beta, and gamma emission, you should compare background to the highest count, and then estimate the drop in count when testing through paper, one-eighth-inch aluminum, and GM tube sidewall.

A positive radiation screen alerts the technician to a hazard that has yet to be described in detail. If a container has been opened and radiation is detected, immediately close the container, retreat to a safe area as defined by your monitor, and take time to make a plan. You should utilize time, distance, and shielding for protection. Decontamination of gloves and other personal protective equipment should occur as well as a shower if the source was in a form such as a powder or gas that escaped the container. Local procedures for a radiation source release including medical monitoring may need to be activated. In the United States any unsecured radiological material falls under federal jurisdiction.

This radiation screen is sensitive enough to detect common household radiation sources such as those sometimes found in a smoke detector, lantern mantle, radium watch dial, orange (uranium) Fiestaware™ pottery, and rock collections. More significant hazards become more obvious when using this method. If the technician elects to continue analysis, a multichannel analyzer based on the type of radiation will help determine the compound and emissions. Once the compound is identified, other chemical and physical hazards may be determined through references.

The radiation type and rate should be determined in order to quantify the radiation hazard. Table 4.4 shows relative dose rates and effects that may be compared to sample values.

Table 4.4 Biological Effects from Short-Term Exposure

Equivalent Dose	Effects
50 Sv or 5000 rem	Immediate illness and death within hours.
10 Sv or 1000 rem	Immediate illness and subsequent death within a few weeks. 1000 rem is the LD100 for humans.
2-10 Sv or 200–1000 rem	Severe radiation sickness with increasing likelihood of death.
1 Sv (1000 mSv) or 100 rem	Threshold for causing immediate radiation sickness in a person of average physical attributes, but unlikely to cause death. Severity of illness increases with dose.
100 mSv or 10 rem	Probability of cancer (rather than the severity of radiation sickness) increases with dose. The estimated risk of fatal cancer is five of every one hundred persons exposed to a dose of 1000 mSv (i.e. if the normal incidence of fatal cancer were 25%, this dose would increase it to 30%).
50 mSv or 5 rem	Lowest dose at which there is any evidence of cancer being caused in adults and the highest dose that is allowed by regulation in any one year of occupational exposure.

Identification of Radiological Material

A radiation expert should be consulted to provide guidance on how to continue analysis for chemical and physical hazards while providing protection against a radiation hazard. Do not discount other chemical and physical hazards in the presence of a positive result from the radiation screen. A radionuclide identifier can identify hundreds of individual radionuclides, but you must consider other hazards of the sample before formulating a work plan. A radionuclide identifier such as the Thermo Scientific RadEye SPRD-H® or FLIR indentiFINDER® series can measure the spectrum produced by an ionizing radiation source and compare it to a library of spectra stored in the device. Spectra are defined by decay characteristics of spin and parity, gamma peak energy, alpha and beta transitions, etc. The ability of the identifier to determine the identity of the source depends on the type and range of sensors in the device as well as the presence of a matching spectrum.

Explosive Material

There is no one method to identify all explosives. Various technologies can be used to identify certain explosives and even screening can be difficult without some presumption of explosive material. Based on your situation, you might try an identifying technology first if there is some indication the material is explosive. Otherwise, you could discover the material to be explosive while using another characterization strategy. In either case, it is important to realize that at the moment you suspect the material may be part of a bomb or explosive device it becomes a law enforcement matter. If you believe you are working with an explosive device, stop. Don't even think about proceeding without notifying the proper law enforcement agency. If you are authorized to characterize or identify explosive material, you should consider using multiple forms of technology to confirm your results, similar to characterization of any other unknown substance.

Characterization of Explosive Material

Explosive material can take on several manners of appearance. Explosives can be solids or liquids, brightly colored or dull and drab, organic or inorganic. These materials are often referred to as "energetic materials," so a big clue to their presence is a material reacting energetically. For example, if in previous testing you placed a small amount in a flame and the material exploded or burned quickly and intensely like a flare, you should consider changing your characterization strategy to focus on explosive material.

A material can be characterized as a possible explosive if it is:

- Shock sensitive
- Thermally sensitive
- A combination of fuel and oxidizer
- Contains nitrate
- Contains perchlorate
- Contains peroxide
- Contains any combination of groups listed in Table 4.5
- Produces a positive result from a commercially available colorimetric explosive screening test

Most of these traits can be found in the water and flame tests in the section on *Characterizing Unknown Material*. The oxidizer tests in water detect most oxidizers, but not substances with organic nitrate and nitro groups. The flame test detects thermally unstable explosives. Some explosives are shock-sensitive and they are detected by the hammer and anvil test, which is about as simple to perform as it sounds; however, this test must be performed safely, as described in Chapter 3.

To determine if a substance is shock sensitive, place a milligram of sample (equivalent to 5 to 10 grains of table salt) on a hard, inert surface and give it a good whack with a hammering device. Since you are suspecting a crisp and powerful explosion, use material for the hammer and anvil that will not fragment when an explosive wave passes through it. Perform this test in a manner that controls any material that may be ejected when struck. Preferably, you should use a device that you can operate remotely so you are removed from the area of any possible fragmentation. Wear appropriate protective equipment such as a face shield and ear protection.

Table 4.5   Characteristic Composition of Explosives

Explosive	Hydrocarbon	Nitro	Organic Nitrate	Nitrate Anion	Perchlorate	Chlorate	Peroxide	Liquid Oxygen	Oxygen	Metal	Picrate	Unstable Structure
Acetylides of heavy metals										•		•
Aluminum containing polymeric propellant	•									•		
Aluminum ophorite explosive					•					•		
Amatex	•	•		•								
Amatol	•	•		•								
Amm nitrate exp. mixtures (cap sensitive)				•								
Amm nitrate exp. mixtures (non–cap sensitive)				•								
Amm perchlorate composite propellant	•				•							
Amm perchlorate explosive mixtures	•				•							
Amm salt lattice w/substituted inorganic salts												•
Ammonal	•	•		•						•		
Ammonium picrate	•	•									•	
ANFO	•			•								
Aromatic nitro compound explosive mixtures	•	•										
Azide explosives												•
Baranol	•	•		•						•		
Baratol	•	•		•								
BEAF	•	•							•			
Black powder	•			•								
Black powder based explosive mixtures	•			•								
Blasting agents, nitro-carbo-nitrates	•	•		•								
BTNEC	•	•										
BTNEN	•	•										
BTTN	•		•	•								
Butyl tetryl	•	•										
Calcium nitrate explosive mixture	•			•								

(Continued)

Table 4.5 (Continued) Characteristic Composition of Explosives

Explosive	Hydrocarbon	Nitro	Organic Nitrate	Nitrate Anion	Perchlorate	Chlorate	Peroxide	Liquid Oxygen	Oxygen	Metal	Picrate	Unstable Structure
Cellulose hexanitrate explosive mixture	•											
Chlorate explosive mixtures	•					•						
Composition A	•	•										
Composition B	•	•										
Composition C A61	•	•										
Copper acetylide										•		•
Cyanuric triazide												•
Cyclonite	•	•										
Cyclotetramethylenetetranitramine	•	•										
Cyclotol	•	•										
Cyclotrimethylenetrinitramine	•	•										
DATB	•	•										
DDNP	•	•										
DEGDN	•		•									
Dimethylol dimethyl methane dinitrate composition	•		•									
Dinitroethyleneurea	•	•										
Dinitroglycerine	•	•										
Dinitrophenol	•	•										
Dinitrophenolates	•	•										
Dinitrophenyl hydrazine	•	•										
Dinitroresorcinol	•	•										
Dinitrotoluene-sodium nitrate explosive mixtures	•	•	•									
DIPAM	•											
Dipicryl sulfone	•	•										
Dipicrylamine	•	•										
DNPA	•	•										
DNPD	•	•										
Dynamite	•	•										
EDDN	•											
EDNA	•	•										

(Continued)

Table 4.5 (Continued) Characteristic Composition of Explosives

Explosive	Hydrocarbon	Nitro	Organic Nitrate	Nitrate Anion	Perchlorate	Chlorate	Peroxide	Liquid Oxygen	Oxygen	Metal	Picrate	Unstable Structure
Ednatol	•	•										
EDNP	•	•										
EGDN	•		•									
Erythritol tetranitrate explosives	•											
Esters of nitro-substituted alcohols	•	•										
Ethyl tetryl	•											
Exp. mix containing sensitized nitromethane	•	•										
Exp. mix containing tetranitromethane (nitroform)	•	•										
Exp. mix containing tetranitromethane (nitroform)	•	•										
Exp. mix of oxy-salts and hydrocarbons	•											
Exp. mix of oxy-salts and nitro bodies		•										
Exp. mix of oxy-salts and water-insoluble fuels	•											
Exp. mix of oxy-salts and water-soluble fuels	•											
Explosive mix containing sensitized nitromethane	•	•										
Explosive nitro compounds of aromatics	•	•										
Explosive organic nitrate mixtures	•			•								
Flash powder*										•		
Fulminate of mercury	•									•		
Fulminate of silver										•		
Fulminating gold										•		
Fulminating mercury										•		
Fulminating platinum										•		
Fulminating silver										•		
Gelatinized nitrocellulose	•	•										
Gem-dinitro aliphatic explosive mixtures	•	•										

(Continued)

Table 4.5 (Continued) Characteristic Composition of Explosives

Explosive	Hydrocarbon	Nitro	Organic Nitrate	Nitrate Anion	Perchlorate	Chlorate	Peroxide	Liquid Oxygen	Oxygen	Metal	Picrate	Unstable Structure
Guanyl nitrosamino guanyl tetrazene	•	•										
Guanyl nitrosamino guanylidene hydrazine	•	•										
Guncotton	•	•										
Heavy metal azides	•									•		•
Hexanite	•	•										
Hexanitrodiphenylamine	•	•										
Hexanitrostilbene	•	•										
Hexogen	•	•										
Hexogene or octogene and nitrated N-methylaniline	•	•										
Hexolites	•	•										
HMTD	•						•					
HMX	•	•										
HNIW	•		•									
Hydrazinium nitrate/hydrazine/ aluminum exp. system				•						•		
Hydrazoic acid												•
KDNBF	•	•										
Lead azide										•		•
Lead mannite	•									•		
Lead mononitroresorcinate	•	•								•		
Lead picrate		•								•	•	
Lead salts, explosive										•		
Lead styphnate	•	•								•		
Liquid nitrated polyol and trimethylolethane	•											
Liquid oxygen explosives	•							•				
Magnesium ophorite explosives					•					•		
Mannitol hexanitrate	•	•										
MDNP	•	•										
MEAN	•											
Mercuric fulminate										•		

(Continued)

Table 4.5 (Continued)   Characteristic Composition of Explosives

Explosive	Hydrocarbon	Nitro	Organic Nitrate	Nitrate Anion	Perchlorate	Chlorate	Peroxide	Liquid Oxygen	Oxygen	Metal	Picrate	Unstable Structure
Mercury oxalate										•		•
Mercury tartrate	•									•		
Methyl nitrate	•		•									
Metriol trinitrate	•		•									
Minol-2	•	•		•						•		
MMAN+A143	•	•		•								
Mononitrotoluene-nitroglycerin mixture	•	•										
NIBTN	•	•										
Nitrate explosive mixtures	•			•								
Nitrate sensitized with gelled nitroparaffin	•	•		•								
Nitrated carbohydrate explosive	•	•										
Nitrated glucoside explosive	•	•										
Nitrated polyhydric alcohol explosives	•	•										
Nitric acid and a nitro aromatic explosive	•	•		•								
Nitric acid and carboxylic fuel explosive	•			•								
Nitric acid explosive mixtures	•			•								•
Nitro aromatic explosive mixtures	•	•										
Nitro compounds of furane explosive mixtures	•	•										
Nitrocellulose explosive	•	•										
Nitroderivative of urea explosive mixture	•		•									
Nitrogelatin explosive	•	•										
Nitrogen trichloride												•
Nitrogen tri-iodide												•
Nitroglycerine	•	•										
Nitroglycide	•	•										
Nitroglycol	•		•									
Nitroguanidine explosives	•	•										

(Continued)

Table 4.5 (Continued)   Characteristic Composition of Explosives

Explosive	Hydrocarbon	Nitro	Organic Nitrate	Nitrate Anion	Perchlorate	Chlorate	Peroxide	Liquid Oxygen	Oxygen	Metal	Picrate	Unstable Structure
Nitronium perchlorate propellant mixtures					•							•
Nitroparaffins (explosive)	•	•		•								
Nitrostarch	•	•										
Nitro-substituted carboxylic acids	•	•										
Nitrourea	•	•										
Octogen	•	•										
Octol	•	•										
Organic amine nitrates	•		•									
PBX	•		•									
Penthrinite composition	•	•										
Pentolite	•	•										
Perchlorate explosive mixtures					•							
Peroxide based explosive mixtures							•					
PETN	•		•									
Picramic acid and its salts	•	•										
Picramide	•	•										
Picrate explosives	•	•										
Picrate of potassium explosive mixtures	•	•								•		
Picratol	•	•										
Picric acid (manufactured as an explosive)	•	•										
Picryl chloride	•	•										
Picryl fluoride	•	•										
PLX	•	•										
Polynitro aliphatic compounds	•	•										
Polyolpolynitrate-nitrocellulose explosive gels	•	•										
Potassium chlorate and lead sulfocyanate explosive						•				•		
Potassium nitrate explosive mixtures	•			•								
Potassium nitroaminotetrazole		•										•

(Continued)

Table 4.5 (Continued) Characteristic Composition of Explosives

Explosive	Hydrocarbon	Nitro	Organic Nitrate	Nitrate Anion	Perchlorate	Chlorate	Peroxide	Liquid Oxygen	Oxygen	Metal	Picrate	Unstable Structure
PYX	•	•										
RDX	•	•										
Salts of organic amino sulfonic acid exp. mixture	•											
Silver acetylide										•		•
Silver azide										•		•
Silver fulminate										•		
Silver oxalate explosive mixtures										•		•
Silver styphnate	•	•								•		
Silver tartrate explosive mixtures	•									•		
Silver tetrazene										•		•
Smokeless powder	•	•										
Sodatol	•	•				•						
Sodium amatol	•	•		•								
Sodium azide explosive mixture												•
Sodium dinitro-ortho-cresolate	•	•										
Sodium nitrate explosive mixtures				•								
Sodium nitrate-potassium nitrate explosive mixture				•								
Sodium picramate	•	•										
Styphnic acid explosives	•	•										
T4	•	•										
Tacot	•	•										•
TATB	•	•										
TATP	•						•					
TEGDN	•		•									
Tetranitro carb azole	•	•										
Tetrazene	•		•									•
Tetryl	•	•										
Tetrytol	•	•										
TMETN	•	•										
TNEF	•	•										
TNEF	•	•										

(Continued)

Table 4.5 (Continued)   Characteristic Composition of Explosives

Explosive	Hydrocarbon	Nitro	Organic Nitrate	Nitrate Anion	Perchlorate	Chlorate	Peroxide	Liquid Oxygen	Oxygen	Metal	Picrate	Unstable Structure
TNEOC	•	•										
TNEOF	•											
TNT	•	•										
Torpex	•	•								•		
Tridite	•	•										
Trimethylol ethyl methane trinitrate composition	•	•										
Trimethylolethane trinitrate-nitrocellulose	•	•										
Trimonite	•	•										
Trinitroanisole	•		•									
Trinitrobenzene	•	•										
Trinitrobenzoic acid	•	•										
Trinitrocresol	•	•										
Trinitro-meta-cresol		•										
Trinitronaphthalene	•	•										
Trinitrophenetol	•	•										
Trinitrophloroglucinol	•	•										
Trinitroresorcinol	•	•										
Tritonal	•	•								•		
Urea nitrate	•			•								
Water-in-oil emulsion explosive compositions	•											

*Flash powder was produced by several manufacturers. Flash powder may contain aluminum, magnesium and/or zirconium as the metal fuel. The oxidizer may be barium nitrate, sodium nitrate, potassium nitrate, potassium chlorate and/or potassium perchlorate. Magnesium carbonate was a common additive.

If the first test is negative for shock-sensitive materials, work your way up to a larger volume, never exceeding a gram (the size of an M&M candy). The larger the sample, the farther away you should be from it. A remote hammer and anvil device will greatly improve safety. Not all explosives are shock-sensitive, so a negative result does not rule out explosives. Some explosives such as plastic C-4 can be safely pounded into shape and when ignited will burn gently; but if shocked while burning, a violent explosion will occur. A gram of exploding C-4 will most likely cause injury. If you suspect a material such as C-4, you should use a technology that does not require simultaneous shocking and burning because a relatively large sample will be required and the test itself would present a hazard.

Unfortunately, there is no good colorimetric test for the universal detection of nitro or organic nitrate groups. Some colorimetric indicators have been used in the past, but have been found to be unstable for field use, toxic or carcinogenic. Some colorimetric test kits are available commercially. These kits will detect many of the more common commercial explosive materials based on organic structures, functional groups, and degradation products. These are very useful in presumptive testing such as post-blast forensics, but they may also produce false-positives in the case of unknown material. These kits are useful in the detection of peroxide-based and chlorate-based improvised explosive material.

Identification of Explosive Material

Some explosive materials can be identified in very small amounts. IMS technology is particularly useful in the identification of explosives as evidenced by the deployment of desktop IMS units in airports. However, some explosives cannot be detected by IMS, and those that can be detected often require more than one test since most IMS instruments cannot operate across the entire range of parameters necessary to detect all explosives within its capability. When operated as designed, IMS instruments are capable of detecting a few nanograms of invisible explosive particles. A 50 micron particle of RDX, which is barely visible with unaided vision, weighs about 100 ng, more than enough to be reliably detected.

Portable IMS units designed for detection of trace explosives have several advantages over other technologies. They are fast, relatively inexpensive, are adaptable to changing environments, and can be used by operators without extensive training. IMS units are limited in the detection of explosives in that they often cannot detect the slight amount of vapor emitting from explosives that typically have very low vapor pressure, but when characterizing a visible sample of suspected explosive, the instrument can be switched to particle mode.

IMS is not as definitive for identification of explosive compounds as some other technologies, but the other technologies are often not available or not practical for field application. False-positives are possible from common, nonhazardous material. False-negatives are possible from interfering material that may be preferentially ionized; these masking agents, whether present intentionally or coincidentally, prevent the explosive compound from being ionized. Other factors may be present in individual cases that prevent accurate analysis of explosive material for which the IMS instrument was designed to identify. Identification will depend on matching library spectra resident in the instrument. Additionally, IMS instruments require strict maintenance, calibration, and operating procedures in order to analyze a result with a high level of confidence.

Many IMS instruments are optimized for the detection of explosives RDX, TNT, PETN, and many other commercial explosives with related structures. In order to realize the full analyzing potential of your IMS unit, you will need to understand how to reconfigure the instrument and also understand its limitations. IMS is generally not optimized for detection of gunpowder, most notably black powder. Since sampling in particle mode requires heating of the sample to 180°C (356°F) or higher, care must be taken to test very small amounts to avoid explosive decomposition. A milligram (approximately five grains of table salt) is generally well above the detection limit for many explosives.

Raman spectroscopy can also be used to identify many explosive materials. Raman spectroscopy can detection of explosive materials that are organic or polyatomic inorganics, such as perchlorates and nitrates. Raman provides a less robust response to organic material that is small and highly polar and organic material containing only single bonds, such as aliphatics, sugars, starches, and cellulose; however, these are considerations, not limitations. Identification will depend on matching library spectra resident in the instrument.

A Raman laser can ignite or explode some explosive materials. Dark materials are more susceptible to heating by the laser. Black powder and silver azide are known to be sensitive to the laser; however, black plastic, latex paint, and cardboard can also ignite. Since many explosives are heat sensitive, and since the instrument must contact or be very close to the material, it could be an expensive and dangerous proposition to cause the sample to explode.

Fluorescent material can interfere with Raman instruments to the point of making the test useless. Fluorescence can be caused by brightly colored material (especially blues, greens, and black), some low-quality glass and biological material, among others. Improvements in laser optics, filtering and data analysis software are able to reduce these problems.

Infrared spectroscopy instruments may be helpful in identifying explosive compounds. Along with the limitations described in Chapter 3, identification will depend on matching library spectra resident in the instrument. Water masks low concentration materials that might otherwise be identified. Since the explosive material may be produced in a clandestine manner to include mixtures of undetermined specification, robust mixture analysis is required to determine impure compounds.

Illegal Drugs

Characterization of suspected illegal drugs will almost always be circumstantial. Some samples will develop suspicion and lead you to test. Field drug tests have long been established with law enforcement. Drugs with or without markings may not be what they appear.

Characterization of Illegal Drugs

The most effective method of characterizing a suspected illegal drug is testing with the *Color Test Reagents/Kits for Preliminary Identification of Drugs of Abuse* (National Institute of Justice Standard 0604.01). Testing is straightforward, as indicated in the instructions. If you don't have the prepackaged test, individual tests may be run if you have the reagents as indicated in Chapter 3. Negative results have fairly high confidence and positive results indicate laboratory testing is necessary. Remember that false-positives are possible from many other substances, thus the screen is used based on suspicion and not as a general screen for unknown material.

Identification of Illegal Drugs

Identification of illegal drugs is now often accomplished by field tests involving Raman, FTIR or HPMS instruments. Some of these instruments are designed strictly to identify drugs, precursors and select materials, such as cutting agents. The specificity improves identification by removing spectra from the library that might otherwise reduce the efficiency of the instrument.

For example, a ThermoScientific Gemini analyzer has about 14,000 Raman and 12,000 FTIR library items for general use, while the Thermo Scientific TruNarc has about 400 items in its library. Both instruments contain similar components, but TruNarc is more efficient for Raman anaylsis of illicit drugs.

HPMS (a form of IMS) is able to sample air and swabs and identify trace amounts illicit drugs. MX908 can be used like an air monitor to search for increasing concentration of a specified item as well as test swabs taken from surfaces with trace contamination.

Better identification might be obtained by a laboratory instrument that can separate the sample and identify it, such as a gas chromatograph-mass spectrometer. Identification will depend on the library containing the target drug spectra and the ability of the GC to separate whatever mixture is contained by the sample.

Formerly, IMS has been used by several agencies as the technology of choice for trace detection of illegal drugs. However, some drugs cannot be detected by older IMS instruments, and those that can be detected often require more than one test because most IMS instruments cannot operate across the entire range of parameters necessary to detect all drugs within its capability. Most drugs are detected when the IMS unit is operated in positive ion mode (most explosives are detected in negative ion mode). Additional calibration and desorber temperature is typically different between modes. When operated as designed, IMS instruments are capable of detecting a few nanograms of a drug.

Portable IMS units designed for detection of trace amounts of drugs have several advantages over other technologies. They are fast, relatively inexpensive, are adaptable to changing environments, and can be used by operators without extensive training. IMS units are limited in the detection of drugs in that they often cannot detect the slight amount of vapor emitting from those compounds that have very low vapor pressure, but when characterizing a visible sample of suspected illegal drug, the instrument can be switched to particle mode.

False-positives are possible from common, nonhazardous material. False-negatives are possible from interfering material that may be preferentially ionized; these masking agents, whether present intentionally or coincidentally, prevent the drug compound from being ionized. Other factors may be present in individual cases that prevent accurate analysis of drug material for which the IMS instrument was designed to identify. Identification will depend on matching library spectra resident in the instrument. Additionally, older IMS instruments require strict maintenance, calibration, and operating procedures in order to analyze a result with a high level of confidence.

Summary

The answer to the question "What is that stuff?" can be difficult under the best of conditions. Your concern for a practical answer to that question, an answer that can be applied to life safety issues in a particular situation, is paramount to safe management of hazardous materials. Many situations require identification of hazards through characterization of an unknown substance; identification of the substance may have to wait for detailed laboratory-based techniques.

Workers from many fields can use information in this book to improve safety while performing their jobs. Emergency first responders, crime scene investigators, laboratory technicians, waste site workers, and others all need safe workplaces. This need can be met through straightforward, concise, and practical procedures in many working environments.

The rapid growth in technology will make more tools available for the characterization and identification of unknown materials. As quickly as this book can be revised and published, more instrumentation will become

Figure 4.17 Free resources such as color images of M8 paper challenged by several household chemicals are available at HazardID.com. Take a look for yourself!

commercially available. The effort to characterize hazards will become more effective through integration of new technology into existing strategies for unknown substances, material used for weapons of mass destruction, explosives, or illegal drugs. Hopefully, new technology will continue to simplify the strategies.

The material presented in this book will help you understand workplace hazards, the technology available to identify those hazards, and strategies for using combinations of technology in qualitative analysis. Visit www. hazardID.com for other resources related to this book. (Figure 4.17).

References

2004 Emergency Response Guidebook, United States Department of Transportation, Washington, DC, 2005.

Aarino, P., *Expert Systems in Radiation Source Identification*, www.tkk.fi/Units/AES /projects/radphys/shaman.htm, Helsinki University of Technology, Finland, accessed December 7, 2006.

ANP Technologies, Inc. website., www.anptinc.com, Newark, DE, accessed December 4, 2006.

Armour, M., Browne, L., and Weir, G., *Hazardous Laboratory Chemicals Disposal Guide*, CRC Press, Boca Raton, FL, 1991.

Belgrader, P., et al., "PCR Detection of Bacteria in Seven Minutes," *Science* 284: 449–50 (1999).

Belgrader, P., et al., "Rapid Pathogen Detection Using a Microchip PCR Array Instrument," *Clinical Chemistry* 44(10): 2191–94 (1998).

Bradford, M. M., "A Rapid and Sensitive Method for the Quantitation of Microgram Quantities of Protein Utilizing the Principle of Protein-Dye Binding," *Analytical Biochemistry* 72: 248 (1976).

Bravata, D., et al., *Bioterrorism Preparedness and Response: Use of Information Technologies and Decision Support Systems, Evidence Report/Technology Assessment No. 59*, AHRQ Publication No. 02-E028, Agency for Healthcare Research and Quality, Rockville, MD, June 2002.

Bravata, D., et al., *Detection and Diagnostic Decision Support Systems for Bioterrorism Response*, www.cdc.gov/ncidod/EID/vol10no1/03-0243.htm, Emerging Infectious Diseases, January 2004.

Drager-Tube/CMS Handbook, 13th ed. Drager Safety AG & Co. KGaA, Lubeck, Germany, 2004.

Eisenbud, M. and Gesell, T., *Journal of Environmental Radioactivity from Natural, Industrial, and Military Sources*, 4th ed., Academic Press, Inc., San Diego, CA, 1997, 135.

Ellison, D.H., *Handbook of Chemical and Biological Warfare Agents*, CRC Press, Boca Raton, FL, 1999.

Engelder, C., Dunkelberger, T., and Schiller, W., *Semi-Micro Qualitative Analysis*, John Wiley and Sons, New York, 1936.

Everett, K. and Graf, F.A., Jr., *Handling Perchloric Acid and Perchlorates*, CRC Handbook of Laboratory Safety, 2nd ed., Steere, N.V., Ed., CRC Press, Boca Raton, FL, January 1971.

Fatah, A., et al., *Guide for the Selection of Chemical Agent and Toxic Industrial Material Detection Equipment for First Responders, Second Edition, Guide 100-04*, Volume I, National Institute of Standards and Technology, Gaithersburg, MD, March 2005.

Fatah, A., et al., *Guide for the Selection of Chemical Agent and Toxic Industrial Material Detection Equipment for First Responders, Second Edition, Guide 100-04*, Volume II, National Institute of Standards and Technology, Gaithersburg, MD, March 2005.

Fiegl, F., *Laboratory Manual of Spot Tests*, Academic Press, Inc., Publishers, New York, 1943.

Fiegl, F., *Spot Tests in Inorganic Analysis*, 5th ed., Elsevier Publishing Company, New York, 1958.

Field Forensics, Inc. website., fieldforensics.com/index.htm, Largo, FL, accessed March 1, 2007.

Gastec Corporation, *Environmental Analysis Technology Handbook*, 5th ed., Ayasishi, Japan, October 2004.

Goodson, L.H., et al., *Stabilization of Cholinesterase, Detector Kit Using Stabilized Cholinesterase, and Methods of Making and Using the Same*, United States Patent 4,324,858, April 13, 1982.

Guidelines for Responding to a Chemical Weapons Incident, Publication L126/QRG-C, Domestic Preparedness Program, Washington, DC, August 1, 2003.

How Can You Detect Radiation? Health Physics Society, hps.org/publicinfor-mation/ate/faqs/radiationdetection.html, accessed August 29, 2005.

James, R, et al., *Industrial Test Systems, Inc. Cyanide Reagent Strip Test Kit*, ETV Advanced Monitoring Systems Center, Battelle, Columbus, OH, April 2005.

Jennings, M., *HHS and USDA Select Agents and Toxins*, Centers for Disease and Control, Department of Health and Human Services, Washington, DC, February 23, 2006.

Jermain, John and Waranauskas, Jennifer, *Explosive Photogenesis – Vintage Photographic Flash Powder*, Detonator, International Association of Bomb Technicians and Investigators, 1120 International Parkway, Fredricksburg, VA, May/June 2015.

Laboratory Safety Guide, University of Wisconsin-Madison Safety Department, Madison, WI, 2006.

Macherey-Nagel, Inc. website., www.macherey-nagel.com, Easton, PA, accessed January 4, 2007.

Medical Management of Chemical Casualties Handbook, 2nd ed., United States Army Medical Research Institute of Chemical Defense, Aberdeen Proving Ground, Aberdeen Proving Ground, MD, September 1995.

Merck KGaA, *Applications for Merckoquant" Tests*, photometry.merck.de/servlet/PB/menu/1170820/index.html, Darmstadt, Germany, accessed January 5, 2007.

Mirzayanov, V., *State Secrets, An Insider's Chronicle of the Russian Chemical Weapons Program*, Outskirts Press, Inc., Denver, CO, 2009.

Mistral Group website., www.mistralgroup.com/default.asp, Bethesda, MD, accessed March 1, 2007.

Mueller, R. and Yingling, V., *Aqueous Film-Forming Foam (AFFF)*, Interstate Technology & Regulatory Council (ITRC), Washington, DC, October 2018.

National Institute for Occupational Safety and Health, *NIOSH Pocket Guide to Chemical Hazards*, Centers for Disease Control and Prevention, Washington, DC, September 2005.

Ong, Kwok Y., et al., *Domestic Preparedness Program: Evaluation of the TravelIR HCI™ HazMat Chemical Identifier*, Research and Technology Directorate, Soldier and Biological Chemical Command, Aberdeen Proving Ground, Aberdeen, MD, August 2003.

Peroxide Forming Chemicals Management and Assessment Guidelines, University of Washington, Environmental Health and Safety, Seattle, WA, 2005.

Pibida, L., et. al., *Results of Test and Evaluation of Commercially Available Radionuclide Identifiers for the Department of Homeland Security, Version 1.3*, National Institute of Standards and Technology, Gaithersburg, MD, May 2005.

Pibida, L., Karam, L. and Unterweger, M., *Results of Test and Evaluation of Commercially Available Survey Meters for the Department of Homeland Security*, Version 1.3, National Institute of Standards and Technology, Gaithersburg, MD, May 2005.

Pibida, L., Karam, L. and Unterweger, M., *Results of Test and Evaluation of Commercially Available Personal Alarming Radiation Detectors and Pagers for the Department of Homeland Security*, Version 1.3, National Institute of Standards and Technology, Gaithersburg, MD, May 2005.

Precision Labs UK company website., www.precisionlabs.co.uk, King's Lynn, United Kingdom, accessed January 4, 2007.

Primer on Spontaneous Heating and Pyrophoricity, United States Department of Energy, Washington, DC, December 1, 1994.

Radionuclides, US Environmental Protection Agency, Washington, DC, December 14, 2005.

Safe Handling of Alkali Metals and Their Reactive Compounds, Environmental Safety and Health Manual, Document 14.7, Lawrence Livermore National Laboratory, Livermore, CA, revised October 13, 2005.

Safety Regulations for Synthetic, Biological, and Physical Chemists, Research School of Chemistry, Australian National University, Canberra, Australia, 2006.

Santillán, J., *FirstDefender Training Kit*, Ahura Corporation, Wilmington, MA, November, 2006.

Sensidyne, Inc., *Sensidyne Gas Detector Tube Handbook*, Clearwater, FL, undated.

Shriner, R. and Fuson, R., *The Systematic Identification of Organic Compounds—A Laboratory Manual*, John Wiley & Sons, Inc., New York, 1935.

Smith, E. and Dent, G., *Modern Raman Spectroscopy—A Practical Approach*, John Wiley & Sons, Ltd., West Sussex, England, May 2006.

Smith, P.K., et al., "Measurement of Protein Using Bicinchoninic Acid," *Analytical Biochemistry* 150: 76–85 (1985).

Stuempfle, A., et al., *International Task Force 25. Hazard from Industrial Chemicals Final Report*, Edgewood Research Development and Engineering Center, Aberdeen Proving Ground, Aberdeen, MD, April 1998.

Sun, Yin and Ong, Kwok Y., *Detection Technologies for Chemical Warfare Agents and Toxic Vapors*, CRC Press, Boca Raton, FL, 2005.

U.S. Department of Justice, *An Introduction to Biological Agent Detection Equipment for Emergency First Responders, NIJ Guide 101-00*, Office of Justice Programs, National Institute of Justice, December 2001.

U.S. Department of Justice, *Color Test Reagents/Kits for Preliminary Identification of Drugs of Abuse*, NIJ Standard-0604.01, National Institute of Justice, National Law Enforcement and Corrections Technology Center, Rockville, MD, July 2000.

Uranium Information Center Ltd website., www.uic.com.au/ral.htm, Melbourne, Australia, July 2002.

Verkouteren, J., et. al., *IMS-Based Trace Explosives Detectors for First Responders*, NISTIR 7240, National Institute of Standards and Technology, U.S. Department of Commerce, Washington, DC, January 2005.

Wayne, C., *A Simple Qualitative Detection Test for Perchlorate Contamination in Hoods*, www.orcbs.msu.edu/chemical/resources_links/contamhoods.htm, Office of Radiation, Chemical and Biological Safety, Michigan State University, East Lansing, MI, accessed January 6, 2006.

Welcher, F.J., *Organic Analytical Reagents*, D. Van Nostrand, New York, NY, 1948.

Techniques

5

Introduction

Field instruments are designed to identify materials with little or no user interpretation. When that happens, everyone is happy. But when that doesn't happen, the need to identify or characterize unknown material remains. This chapter contains techniques to manipulate samples to make field testing more successful.

Manipulating samples carries more risk. Very small sample sizes will decrease exposure risk. Always follow guidance from your employer and instrument manufacturer.

Separating Sample Components

Field samples are rarely pure, and any field sample may present testing challenges for various technologies. The presence of several ingredients may overwhelm an instrument's ability to identify some or all components. A dark or fluorescent component can mask Raman-responsive materials. Nonhomogenous samples and very hard substances can present challenges to FTIR analysis. Low-vapor-pressure solids and liquids may not produce vapors for analysis by HMPS.

This section will describe several manipulations an operator may use to improve field analysis. A simple, low-cost tool kit is described that will enable these sample manipulations.

Liquid/Liquid Mixture Separation

Liquid mixtures may be separated if components have different volatility, solubility or molecular weight.

A lighter liquid can be driven from a heavier liquid with gentle heating. The vapor may be tested with HPMS and the remaining heavier liquid will be concentrated to favor identification by FTIR or Raman.

Solvent extraction might be helpful for field testing, especially when strongly polar and nonpolar liquids solvents can be used against each other to draw a partially soluble liquid from one to the other. Water is a strong choice for Raman analysis since it will not produce a Raman response that would compete with the target material. For example, an alcohol in dark, waste motor oil can be extracted by adding water. The alcohol can be identified in the water by Raman, if the concentration is adequate.

Heavier liquid fractions will sink to the bottom, floating the lighter portions above. When first observing a sample, resist the urge to shake it! Simpler mixtures in layers are much easier for your instrument to sort out than one complex mixture.

A coliwasa or stinger tube can be adapted to any liquid container. The stinger tube, open top and bottom, is gently pushed through the liquid from top to bottom. The top end is sealed with a gloved thumb and the stinger tube is removed to reveal a layered aliquot. If the tube is clear, a Raman device can be used to test each layer. Layers can be separated by placing the lower end over a clean container and controlling the release of each layer by controlling the upper opening. Use a narrow diameter stinger for thin liquids and a wider diameter stinger for thicker liquids.

Evaporation

Evaporation can be used to eliminate lighter fractions of a liquid sample, leaving and concentrating the heavier fractions. Evaporation can be passive if a high vapor pressure solvent is present. Consider a 50% mixture of acetone in mineral oil. The acetone will thin the liquid so that a drop placed on a watch glass will spread out quickly. Watch the edges over a minute or so. If the diameter of the liquid on glass begins to shrink, a volatile sample is evaporating and concentrating the oil. Without the acetone, an instrument will be better able to identify the oil without the acetone producing signal that might overwhelm the weaker oil signal.

Evaporation can be actively applied through the use of heat. Moderate heat can be supplied by a butane lighter. More robust heat is produced with a propane torch with a low flame. Place 1 ml (1/2 inch or 10 mm) of liquid in a test tube and hold with a test tube clamp. Heat gently with the propane torch by passing the flame over the base of the test tube with a fanning motion. Keep the test tube pointed in a safe direction in case contents are ejected. Ensure adequate ventilation of toxic or flammable vapors.

As the liquid is heated, lighter fractions will evaporate or boil and move up the neck of the test tube. Initially, some vapor may condense on the upper portion of the test tube until the neck of the tube gets hot. If the condensation is to be tested, stop heating and test a sample. Continue heating the liquid in the base until very little is left and then test this concentrated, heaviest faction.

Finally, continue gentle heating until all the liquid is gone. Any remaining solid that may have been dissolved in the liquid may be tested. Avoid heating the solid any more than is necessary to remove the liquid. Once dry, the solid will heat up very quickly and may degrade.

Do not inject heated vapor or smoke directly into HPMS. Instead, quickly wave an open plastic bag to capture and cool a very dilute sample.

And remember, hot glass looks just like cold glass.

Layers

A liquid sample separated in layers presents an opportunity to test each layer separately – do not shake! If two layers are present in a sample, most have a water-based lower layer and a nonpolar upper layer. If three layers are present, the upper two layers are likely the same as above with the third, lower layer containing a chlorinated solvent. There may be several components to each layer and there may be some crossover between layers due to partial solubility.

Test a layered liquid sample from the top down. Use a pipet to draw a small amount off the top layer for testing and then gently pour off the top layer or use a pipet to draw up the top layer and transfer the remainder to another container and label it.

Expect a film of the top layer to remain on the next, lower layer. Insert a pipet through the film into the second layer while gently squeezing the bulb of the pipet. The top film will be prevented from entering the pipet stem as the air is expelled. Draw a half-bulb full of liquid from this second layer and as the pipet is withdrawn through the film on top, gently squeeze a little liquid out to prevent the film layer from entering the pipet stem. Use some absorbent paper to wipe any film layer off the outside of the pipet stem. Film contaminant on the outside of the stem can flow around a drop being squeezed out and the film may be positioned between the drop and FTIR diamond test surface. Squeeze out a drop onto the absorbent paper and then expel the remainder into a new, labeled container for testing.

Repeat this process for a third layer, if present. Each layer may be tested as a liquid/liquid layer as described above. Raman testing may be done through the test tube wall. FTIR testing will require a drop to be withdrawn and protected from film contamination. The same drop may produce vapors identified by HPMS; however, trace amounts from other layers may still be identified by HPMS. Evaporation will reveal any solids suspended in the liquid layers.

Soluble Extraction

Soluble extraction is a way to separate one liquid from another based on relative solubility.

Consider a 50% mixture of ethanol in dilute hydrochloric acid, which also contains a deep blue dye. Raman testing this sample as an unknown material will be complicated by fluorescence caused by the deep blue dye. The dilute hydrochloric acid is not Raman reactive. It is too weak to be identified by FTIR as an acid, and the low molecular weight is not identified in vapor form by HPMS.

Place 1 in of the sample liquid in a test tube and mark the level with a marker. Add an equal amount of water and observe, then gently agitate by gently squeezing and releasing the bulb of a pipet in the test tube. If the sample appears to be uniform (which it would be), the sample is polar. If the sample separates and floated on the water, the sample is nonpolar.

Repeat the previous step using mineral oil instead of water. After gentle agitation, the mineral oil will float and draw some ethanol into it. The dye will favor one or the other layer. Test each layer.

Each layer should be pH tested to determine the presence of corrosives. pH results of an oil layer will produce a neutral pH result.

Each solvent extracted layer may be further manipulated as a liquid/liquid sample as described above.

Paper Chromatography

Paper can be used to separate components of a liquid sample. A drop or two of liquid is placed on the end of a thin strip of paper from a white coffee filter. The components will travel laterally on the paper with the lightest materials moving farthest from the drop placement. The characteristics of the liquid may require a few tries of differing amounts of liquid to get the best separation.

Immediately test differing areas of the paper, preferably with FTIR and secondarily with Raman. FTIR testing may be performed by lightly holding an area of wet paper against the diamond test surface with a gloved finger. Excess pressure will introduce a signal from the paper, which may interfere with the liquid signal. Raman testing may be performed by placing the laser focal point on various areas of the wet paper. Hold the paper in the air so there is no background interference from a table or other background material. Using white rather than brown filter paper will reduce interference from the dark color.

Solid/Liquid Mixture Separation

Physical separation of solids from liquids help the operator test individual components of a mixture, or at least components that have reduced amounts of interfering material.

Filter

A filter may be used to remove a liquid from a solid that remains in particle form. These particles may be part of a slurry, suspension or excess solid in a saturated solution.

Cut a white coffee filter to fit a small mesh strainer. Avoid excess paper above the strainer rim since liquid will wick up the paper, which is more difficult to control. Add some sample to the paper in the strainer and catch the liquid with a small container (Figure 5.1). The liquid may be tested immediately.

Any solid on the paper will be wet. Scoop out a little of the wet solid and dry it. Spreading the wet solid across a surface can be used to air dry the solid and then scrap it into a small pile for testing. Wet solid can also be dried in the base of a test tube with a gentle application of torch heat. Other methods of adding heat may be used here; sunshine, incandescent light bulb, engine exhaust manifold, and so on. When the solid is dry, it may be tested without liquid interference.

Evaporation

Evaporation is a helpful method of isolating a solid that is dissolved in a liquid.

Active or passive methods of evaporation may be used to remove a liquid from a solid; however, the liquid is no longer available for testing by Raman or FTIR instruments. Forced evaporation through the application of heat makes vapors more readily available for HPMS testing, but be careful not to overload the instrument.

Spreading the wet solid across a surface can be used to air dry the solid and then scrape it into a small pile for testing. Wet solid can also be dried in the base of a test tube with a gentle application of torch heat. Other methods of adding heat may be used here; sunshine, incandescent light bulb, engine exhaust manifold, and so on.

Figure 5.1 A coffee filter held by a mesh strainer used to separate insoluble solid material from a liquid.

Sediment

Solids that are suspended in liquid may be concentrated at the bottom of a container with a centrifuge. This is not practical for field use, but allowing a solid to settle to the bottom over time in response to gravity is something that can be practical.

If two solids with differing solubilities are mixed together, they might be separated by the addition of a liquid. If one solid is soluble in the liquid and the other solid is not soluble, the first solid will dissolve into the liquid while the insoluble solid will either float or sink. Liquid and solid portions can then be separated and tested.

Solid/Solid Mixture Separation

Solids that are visibly or invisibly mixed may be separated for field tests by liquid extraction, the use of a screen or sieve or by other physical means.

Liquid Extraction

A liquid may be used as a solvent to separate a soluble solid from and insoluble solid. This is often useful to separate a powdered metal fuel from a water-soluble oxidizer.

For example, mixtures of one or more powdered metals mixed with one or more oxidizers are considered flash powders. Mixing with water will dissolve the oxidizer into the water and when allowed to settle, the metal powder will separate. Each may be tested separately.

Sieve

A sieve or mesh screen can be used to separate solids based on size. Hardware cloth and food strainers provide a range of mesh sizes. Finer particles fall through the sieve and larger particles are sorted based on mesh size.

Select Separation

Solid mixtures of several components can be isolated visually by using tweezers. This method is useful when particles are similar in size but differ visually, such as mixed fertilizer.

Sample Took Kit

These items are required for a sample tool kit:

- Polyethylene pipets, small and large caliber stems
- Scotch brand translucent tape (matte finish)
- Razor blade in retractable holder

- Small container of deionized water
- 4 ml glass vials with lids
- Watch glass or heavy glass dish or ashtray
- Disposable borosilicate test tubes
- Hand-held test tube clamp
- Black Sharpie marker
- Disposable absorbent paper towels
- White coffee basket filters
- Small stainless steel spatula
- Plastic hotel room key or similar plastic form
- Wand-style butane lighter
- Propane torch
- Mesh strainers of various sizes

Remote Sampling with Robotic Tools

Sometimes sampling is too dangerous for humans to accomplish. A site may be too radioactive to expose a human team. Improvised explosive devices, unstable building structure, gunfire or environmental factors could all cause a delay in sampling by humans.

Advanced robotic tools improve human safety is many situations. Robotic sampling prevents contamination and potential damage of expensive equipment.

A proprietary remote sampling system for Telerob robots can be configured for several sampling missions. The system consists of a magazine with up to five sealable metal containers. Four types of containers may be loaded into the magazine based on mission requirements (Figure 5.2). Each container lid doubles as a collection tool. The four types of tools are:

- "Shovel" for scooping solid and semi-solid materials
- "Cotton bud," a large swab used to absorb liquids
- "Swab" to hold single-use swabs compatible with ion scanners
- "Grasp," similar to tweezers to pick up larger samples

The robot has a memory for the position of each container in the magazine. The operator selects a tool and an automated function causes the gripper to free a tool. The operator manipulates the gripper to collect the sample. An automated function then causes the robot to place the tool in the appropriate container and seal it. The robot transports the sample to a safe area for testing or repackaging.

The ChemThief remote sampling tool (Figure 5.3) is an accessory compatible with most robot platforms. It can collect multiphase samples in the

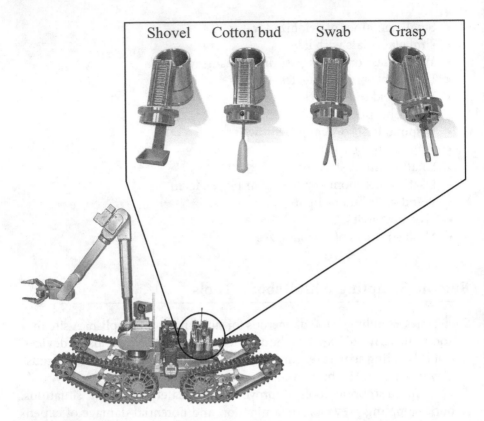

Figure 5.2 A Telerob robot with mounted sample collection tools and containers. (Image: Telerob Gesellschaft für Fernhantierungstechnik mbH. With permission.)

open or by piercing light containers. The tool uses a proprietary design to collect the EPA recommended sample size, a 2 ml or 2 g aliquot. Samples are packaged in a borosilicate glass container and sealed with a nonreactive foam gasket and magnetically sealed lid. The sample is then transported by the robot to a safe area for controlled testing or repackaging. The device is suitable for sampling explosive, radioactive, toxic, flammable, oxidizing or corrosive materials and all CW and BW agents.

Sample container view ports allow an operator to view the 2 ml or 2 g sample before opening the container. Gamma radiation, Raman and HPMS testing may be performed before opening the container (Figure 5.4).

FTIR Techniques

FTIR testing involves placing a sample on a diamond test surface while an emitted infrared beam is reflected back into the instrument. The beam

Figure 5.3 A ChemThief sample collection device in the gripper of a Robotec robot. The lance is being positioned to pierce the container and draw a sample before packaging and transport.

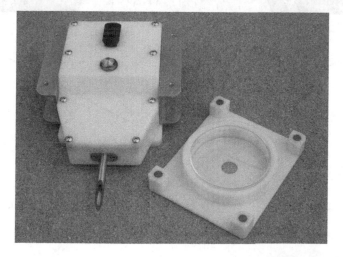

Figure 5.4 ChemThief robotic sampling device. The sample container may be connected by magnets below the sample collection module. Glass view ports allow Raman testing before opening the container.

penetrates into the sample only a micron or two. This method, called total attenuated reflectance (ATR), means FTIR instruments only test the surface of a sample that is in contact with the diamond test surface.

Before wide adoption of ATR, lab-based FTIR instruments required a sample that was sliced 1 μm thick. The sample was then placed between an FTIR beam source and receiver. If the sample was too thick the beam would not penetrate. Field instruments using ATR eliminates the need for precisely sliced samples.

Knowing that a field-portable FTIR instrument can "see" only one or two microns into the edge of the sample will encourage good sample presentation on the diamond surface. Liquids will flow smoothly over the surface while solid materials require pressure to eliminate air gaps and ensure good contact (Figure 5.5).

Preventing the sample from changing in any way will produce the most representative result. For example, evaporation, movement, sediment, contamination, and so on may affect the scan.

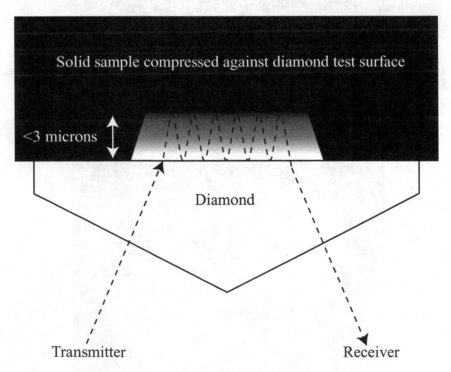

Figure 5.5 The sample must make uniform contact with the diamond test surface to produce an accurate result. Liquids easily flow to make uniform contact. Solids must be compressed against the diamond to displace air gaps.

Water is often used as a solvent. Remember that water has a loud FTIR response and may overwhelm some lesser responsive components. Other liquids may be suitable as solvents, such as alcohols, ketones, and so on.

Nonvolatile Liquid

Nonvolatile liquids are the simplest sample to test with FTIR. When a drop is placed on the diamond the sample makes good contact and does not change. Water will bead up with its high surface tension and is very forgiving. Oils and other nonvolatile liquids simply remain in place for a long time.

A nonvolatile liquid can be tested on a cotton swab. The swab may be used to collect the sample and then placed lightly on the diamond surface. Capillary action will cause the liquid to flow between the cotton and diamond test surface if very light to no pressure is maintained. Ensure that the diamond surface is clean and dry and then run a background scan. When complete, gently touch the saturated cotton swab on the diamond test surface and maintain contact until the scan ends (Figure 5.6).

Volatile Liquid

A volatile liquid may evaporate before the scan is complete, especially in hot, dry conditions. If applying a drop of sample with a pipet, the operator can apply one or more drops as the sample evaporates, but extremely volatile liquids may not remain on the diamond test surface long enough for the instrument to complete a scan.

A small piece of absorbent paper can be used to maintain steady liquid contact on the diamond surface. The capillary action of the volatile liquid will lift the paper above the diamond surface without the paper interfering with the scan.

Figure 5.6 A saturated, or even moist, cotton swab will deliver liquid to the diamond test surface through capillary flow.

Use an absorbent paper, such as kitchen paper towel or hand towel. Cutting a small piece rather than tearing it will avoid crushing the absorbent structure of the paper or inadvertently adding a contaminant from gloves or fingers. Cut or carefully tear the paper to form an approximately 3/8 in (10 mm) square. Ensure that the diamond surface is clean and dry, and then run a background scan. When the background scan is complete, place the dry paper square over the diamond surface. Saturate the paper with liquid sample and start the scan. As the sample evaporates, add more liquid to the point of saturating the paper until the scan ends (Figure 5.7).

Oily Liquid

An oily liquid with its low vapor pressure tests well on the diamond surface. Even a one- or two-micron-thick smear is enough material for a successful scan.

An oily sample can be collected using nonabsorbent material and smeared on the diamond test surface from a glove, plastic or soaked cardboard (Figure 5.8).

The amount of pressure applied to force the sample against the diamond test surface can affect the result when testing a wet, solid sample, such as oil in cardboard. Very light pressure allows the liquid to flow between the solid and diamond surface and the instrument will identify the liquid. High pressure will force the cardboard against the diamond surface while squeezing the liquid away from the test surface, resulting in identification of the solid (cellulose or some other cardboard component). High pressure always favors

Figure 5.7 Volatile liquids may be held on the diamond test surface with a piece of absorbent paper. Do not apply pressure to the paper. Capillary flow will maintain liquid contact with the diamond test surface while floating the paper above the area of the infrared beam.

Figure 5.8 Light pressure from a gloved finger will encourage capillary flow of liquid between the diamond and wet solid material.

identification of the solid while light pressure always favors identification of the liquid.

Liquid Mixture

Liquid mixtures affect the presentation of the sample to the instrument. Homogenous samples of bulk amounts of FTIR-responsive materials should be easily identified.

Generally, field testing two-layer liquids will be found to have a water-based layer on the bottom and a hydrocarbon-based layer on top. Various materials may be dissolved in the upper layer, bottom layer or both in differing concentrations. Three-layer liquids generally have a chlorinated hydrocarbon-based layer on the bottom, a water-based layer in the middle and a hydrocarbon-based layer on top.

Miscible mixtures, such as methylethylketone and water, will produce similar results for both upper and lower layers. Field FTIR instruments are qualitative, not quantitative, identifiers. Some instruments display results ranked by signal strength, not concentration.

In any case, do not shake the sample and mix the layers. The instrument will produce more accurate results if each layer is tested separately rather than attempting to sort all the ingredients of all three layers.

Drawing a sample from a lower layer may contaminate the sample and affect placement on the diamond surface. Collect and test a sample from the upper layer first. Then, consider drawing a sample from the lower layer of a two-layer liquid of oil over water, as shown in Figure 5.9.

Gently squeeze the pipet bulb when inserting the stem so that the upper layer is prevented from entering the stem. Draw the lower liquid into the pipet. As the pipet is withdrawn from the container, slowly squeeze a little of the sample out of the pipet, again to prevent the upper layer liquid

Figure 5.9 Drawing an FTIR sample from the bottom of a two-layer liquid should be accomplished without interfering contamination from the upper layer.

from entering the stem. The outside of the pipet stem will be coated with the upper layer. This light film of oil on the stem of the pipet can coat a drop of the water-based liquid as it is placed on the diamond surface and cause the instrument to identify the oil. This can be mitigated by wiping the stem of the pipet with an absorbent material and expelling a drop or two of liquid before placing a drop of sample on the diamond surface.

Prills and Granules

Prills and granules are small solid structures that may be tested individually by a skilled operator.

Products in prill form may be mixtures, such as fertilizer (Figure 5.10). Each type of prill can produce a different result. Separating each type of prill and testing each will produce a representative result of the total product. Individual prills or granules may require brushing or crushing to remove particles from other components.

Prills may be coated with a water-resistant coating, a layer of degraded product, a second ingredient (Figure 5.11) or contamination from unintentional contact with other materials.

Figure 5.10 Prills or chipped ingredients may be coated with a protective or water-resistant layer. Powder from one component may contaminate another.

Figure 5.11 Shock-sensitive improvised exploding target material consisting of ammonium nitrate prills coated with aluminum powder.

Applying high pressure to the sample on the diamond surface may or may not be successful in forcing the inner portions of the prill through whatever coating might be on the prill. Larger prill samples can be split with a blade. Stick a piece of tape on the shell of the split prill and then use the tape to position the split surface over the diamond test surface. Apply high pressure and scan.

Prills with a coating can be dissolved and the solution tested if the coating is insoluble. Figure 5.12 shows exploding target material, ammonium nitrate prills coated with aluminum powder, in water. After agitation, the solution is

Figure 5.12 Improvised exploding target material, ammonium nitrate coated with aluminum metal powder, is mixed with water and shaken (left). After allowing a short time to settle, the ammonium nitrate dissolved into the water while the aluminum metal separated from the solution. The clear layer may be tested and identified as ammonium nitrate solution.

allowed to settle. The ammonium nitrate dissolves in water and the aluminum powder separates; some sinks and some is collected by surface tension. A pipet is used to remove some of the water solution. A drop placed on the diamond test surface will produce a result for the liquid if no pressure is applied.

Pills

The inner portion of pharmaceutical pills may or may not be homogenous. One clue is the presence of a score mark (Figure 5.13). Pills with a score

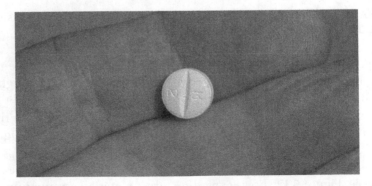

Figure 5.13 A pill with a score mark through the middle of the face indicates the ingredients are uniformly distributed throughout the pill.

mark are homogenous, since the score is intended to provide a half-dose. Pills without a score mark may or may not be homogenous. Some pills are intended to be ingested whole. These pills may contain pockets of different ingredients that are placed in a mill and compressed into a single pill meant to be a single dose. These nonhomogenous pills may produce several results depending on which portion of the pill is placed on the diamond test surface. Dissolving the pill in a solvent can make the mixture homogenous; however, lower signal-producing ingredients may be diluted so that the instrument does not report the result after dilution.

Powder

Powder that is firmly compressed against the diamond test surface should be expected to produce good results for FTIR-responsive materials. Figure 5.14 shows a brown powder containing heroin on the diamond test surface. The substance was field tested with a Raman instrument specific for controlled substances, which produced smoldering and no result.

A powder that has been maliciously dispersed may be difficult to collect in bulk without background contamination. Figure 5.15 depicts a few milligrams of powder, which can be picked up with a piece of transparent tape and positioned over the diamond test surface. Use a tool to concentrate a small amount of powder and then gently press the tape on it to pick it up. Apply only as much pressure as needed to produce a signal. Too much pressure may force fine powders into the tape adhesive so that the adhesive is in contact with the diamond test surface rather than the powder.

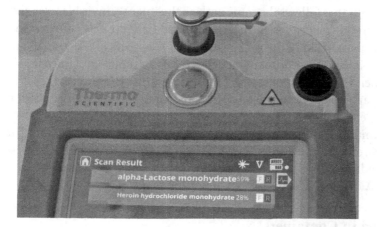

Figure 5.14 A mixture of heroin hydrochloride and alpha-lactose monohydrate, in a form commonly known as brown heroin.

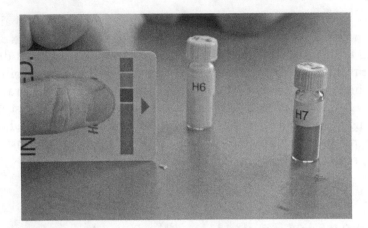

Figure 5.15 A few milligrams of solid material can be scraped into a pile and picked up with a piece of transparent tape. Use the tape to position the small amount of powder on the center of the diamond test surface.

Raman Techniques

The key to Raman field testing is the proper placement of a sample in the laser focal point. The focal point is the area where the sample is bombarded with concentrated laser energy, which produces the most signal available to the instrument.

The intense, focused infrared laser energy used in Raman instruments makes it possible to "see" a few millimeters into a sample compared to the surface analysis using a beam of FTIR energy. A Raman infrared laser is much more powerful than a broad band of infrared light used in an FTIR instrument. A 785 nm laser used in a hand-held Raman instrument produces some visible light that follows the same path as the intense, invisible laser energy that makes up the majority of the beam. This red light is easily displayed on a piece of white paper held in the path of the laser (Figure 5.16). The small amount of red light indicates the focal point when it forms a small and intense red spot on the paper. If the red spot is larger and more diffuse, it indicates that the plane of the paper is either too close or too far from the laser aperture and the laser beam is either converging before or diverging after the focal point.

Raman samples may be tested in a sealed glass vial, in bulk through a transparent or translucent container or as an open spill. Regardless of the test condition, it is important that the operator places the sample in the focal point. A difference of 1 mm can affect the result (Figure 5.17).

Testing in Containers

A Raman laser can penetrate a transparent or lightly colored container if the container wall is thinner than the focal length of the laser. A thicker

Figure 5.16 Hold a piece of white paper in the path of the laser to practice locating the focal point. Move the paper closer and farther from the source to locate the smallest dot of light, which indicates the plane of the paper is at the focal point. Follow manufacturer's guidance for eye safety.

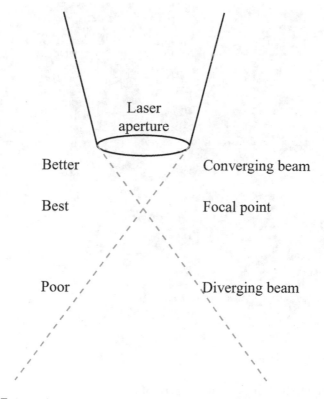

Figure 5.17 Signal return is most intense when the sample is placed in the focal point. Moving a thicker sample closer to cover the entire aperture will produce good signal return by projecting the focal point deeper into the material. Moving the sample farther away from the focal point will diminish the amount of Raman-shifted light that will randomly return to the instrument.

container wall will prevent the operator from extending the focal point into the sample material (Figure 5.18).

Borosilicate glass produces little Raman response to compete with sample return. Light plastic containers produce louder Raman responses that can compete with sample signals. Colored, thick plastic and metal containers will prevent Raman analysis of the contents.

The laser can generate heat (a hazard) or fluorescence, both of which can obscure the instrument's ability to collect and analyze Raman-shifted light necessary to identify the sample.

Heat tends to be generated by darkly colored samples; the darker the color, the more heat is generated. Lightly colored samples may still generate a small amount of heat. Significantly, some explosives and other highly reactive compounds can generate enough heat to react. Therefore, testing bulk samples that cannot be visually assessed in a colored container is not recommended.

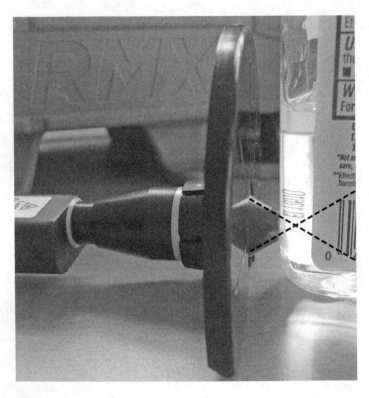

Figure 5.18 Dashed lines show the path of laser energy. The focal point is projected into the contents of a transparent container while the plastic wall remains out of focus. The strongest return signal is generated at the focal point. Moving the probe slightly away from the container will produce more signal from the plastic container and less signal from the contents.

Fluorescence tends to be generated by strongly colored materials, especially toward the green-blue-black end of the visible spectrum. Proteins and lipids also fluoresce and make field Raman instruments a poor choice for biological identification.

An operator will produce the most effective analysis by correctly placing the laser focal point in the sample and applying techniques to minimize the interference of heat and fluorescence.

Some instruments are designed to accept a small glass vial into a holder so that the laser is positioned properly. The vial may hold up to a few grams of sample, yet only the material in the focal point is tested. As a result, only a few milligrams of solid or a couple drops of liquid are needed if positioned in the vail correctly (Figure 5.19). A small sample may be tested outside the vial compartment by placing the sample exactly in the focal point. However, it is more difficult to align the sample visually, the risk of eye injury is increased and ambient light may degrade the Raman spectrum.

Manufacturers may warn against sealing a vial containing a solid, which may be explosive, recommending the vial remains uncapped to prevent pressurization in the event of thermal decomposition. However, other hazards may exist, such as inhalation or contact hazards of dust from a fentanyl derivative, a biological agent, radioactive dust, and so on. Safe operation is the responsibility of the operator. Minimizing the amount of sample will significantly reduce the hazards. Importantly, safety procedures and adequate personal protective equipment will reduce the potential for harm.

Prills and Granules

A prill or small granule may be tested if placed directly in the laser focal point. Testing very small samples may be complicated by the background

Figure 5.19 Two drops of liquid in a vial is adequate volume for a Raman test.

material supporting the sample. To eliminate background signal from sample signal, support the sample in the air with nothing behind it for a few inches. A piece of tape may be used to hold the prill or granule in the focal point. Transparent tape eliminates fluorescence from a colored background. Thin plastic tape produces less signal than thicker plastics, such as fingerprint tape. From practical experience the best tape seems to be Scotch brand matte finish Magic™ Tape. Matte finish tape appears cloudy but produces less competing signal than transparent tape. This particular matte finish tape is cloudy because it contains tiny air bubbles that reduce the density of the plastic film.

Use the tape to pick up the sample and then position the sample in the laser focal point by using the red spot as a reference point (Figure 5.20). Avoid starting the laser until the last moment before scanning the sample. If the sample will not stick to the tape, fold the sample into the tape forming a two-sided envelope.

A small solid sample too large or irregular to be held by tape may be placed in the focal point in one of two ways.

The first is to position the sample on a surface and position the laser above the sample with the focal point in the sample.

Solid samples more than ¼ in (6 mm) thick and not clear will block laser energy from traveling through the sample, reacting with the background and affecting the scan.

If a background material might affect the scan, simply remove the background from laser illumination by suspending the sample in the air with a tool, such as a set of tweezers. Avoid metal tools, which may reflect the laser and cause an eye hazard. Plastic and other materials in tools in the path of

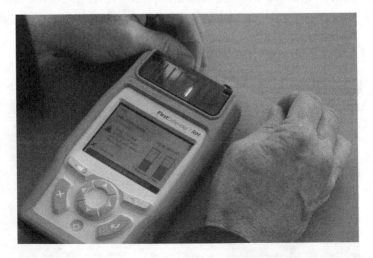

Figure 5.20 A very small sample is held in the laser focal point with tape. The plane of the tape is a few millimeters off the tip of the nosecone.

the laser may be reported with the sample materials in a mixture result. If a result begins with "poly-" it is likely from the plastic tool. This, or any other tool interference, can be verified by scanning the tool with the laser in a separate scan.

Powder

A powder dispersed over an area can be concentrated and then scanned. For example, a fine powder on a tabletop can be scrapped into a small pile using a nonsparking tool, such as a piece of plastic approximating a credit card. Once a pinhead size pile is formed, use a piece of matte finish transparent tape to pick up a thin layer. This thin layer, if positioned exactly in the focal point, can generate enough signal to produce an identification.

Potential Explosive

Scanning potential explosives can be interesting and exciting rather than frightening if the sample size is minimized to a few milligrams. Even samples that react in the laser are not dangerous if a very small amount is used and precautions are taken to prevent a spark from igniting other materials. Gunpowder, black powder, flash powder, and so on are easily ignited by a 785 nm laser, but safe testing still provides information about the identity of the sample by the way it reacts in the laser beam (Figure 5.21).

Very small amounts of explosives may be successfully tested on tape. In fact, a sample library of explosive materials may be enclosed in tape envelopes and labeled for training purposes. Materials such as C4 and other

Figure 5.21 Flash powder residue after laser induced ignition. Note residue on the nosecone.

nonlaser-reactive explosives store well and may be tested safely many times (Figure 5.22).

An explosive ordinance disposal (EOD) group produced what appeared to be an old fire brick (Figure 5.23), one of many stacked in an old bunker built for World War II. The brick was carried on white paper, which caught crumbs as it was carried. The operator safely tested the dark, potential explosive by minimizing the sample size. Some of the crumbs were immobilized on a piece of matte finish transparent tape. The tape was used to position one

Figure 5.22 A few milligrams of commercial or military explosive samples may be scanned through a tape envelope. These samples may be stored and used for practice. From left to right, Semtex 1A, Composition B, Composition C-4, Tetryl, DETA Sheet. All but the DETA Sheet sample produce Raman identification. DETA Sheet fluoresces in a 785 nm laser but is easily identified with FTIR.

Figure 5.23 A sample appearing to be an old fire brick, approximately 1 kg, on white paper.

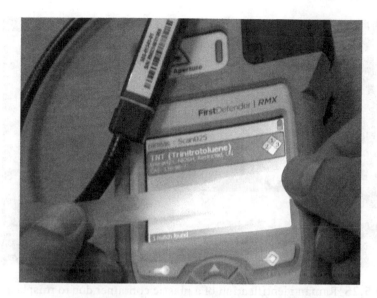

Figure 5.24 A Raman identification of trinitrotoluene resulting from specks of sample suspended on tape in the laser focal point.

crumb in the focal point of a Raman laser set on high power. The instrument quickly produced a result of trinitrotoluene (TNT), as shown in Figure 5.24.

Scanning through Containers

Raman lasers can penetrate container walls and identify the material within.

Significantly, the operator must place the focal point in the contents of the container and not in the wall of the container. Failure to do so will likely result in the identification of the container material and not the contents. Practicing with a transparent container of hand sanitizer and varying the position of the focal point by only a few millimeters will produce results identifying the contents only, contents and container wall or container only.

A Raman result of "poly-" is likely caused by misplacement of the laser focal point in the container wall (Figure 5.25). Try the scan again with the focal point projected into the contents of the container.

Scanning through a container wall that is not transparent could cause an inadvertent reaction if the laser strikes a bulk amount of reactive materials, such as gunpowder.

Coated Samples

Raman instruments are not limited by coated samples, unlike FTIR instruments. The laser can penetrate up to roughly ¼ in (6 mm) into an opaque sample, penetrating the coating. The depth of the sample area proportionally

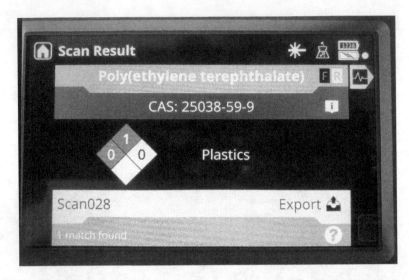

Figure 5.25 Raman identification of a plastic container due to misplaced focal point.

favors signal response to the core of a coated pill. The closer the coating is to the focal point, the more likely it will be reported.

Raman testing allows for nondestructive testing of potential evidence. Figure 5.26 shows pills analyzed in cases of potential abuse. The two pills at the top of the image have consistent color and markings for oxycontin 20 mg pills and are a controlled substance. The pills were tested with a Raman instrument and only alpha-lactose was identified. Knowing that 20 mg of oxycodone hydrochloride should have been identified by a Raman instrument, the pills were analyzed in a laboratory. The pill weight did not match the manufacturer-specified weight. GC-MS found no oxycodone and the pills were found to be counterfeit.

The center pills were unmarked except for the score mark that indicated the ingredients were mixed uniformly throughout the pill. The pills were in a small envelope marked "cough medicine" and an opioid was suspected. Raman testing indicated guaifenesin, an uncontrolled, over-the-counter substance.

The pills at the bottom were marked with the number 377 on a white, elliptical oval tablet, consistent with controlled substance tramadol hydrochloride, 50 mg. Tramadol was confirmed with a Raman instrument.

Leaking Package

A liquid leaking from a cardboard package may be tested with a Raman instrument, with some caveats.

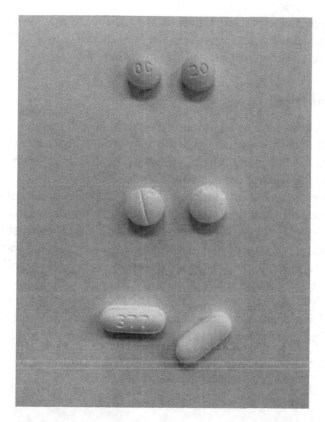

Figure 5.26 Counterfeit oxycontin (top), veterinary grade guaifenesin (middle) and pharmaceutical grade tramadol (bottom).

First, cardboard will produce a Raman response, such as cellulose, which is not strongly Raman responsive. Nonpolar liquids produce a low Raman response, while polar liquids have a strong Raman response. All things being the same, cellulose is more likely to be identified if the liquid is nonpolar. Since water is the only common liquid that will not produce Raman signal in a 785 nm laser, a result of cellulose as the only component of the sample suggests the liquid is water based.

Second, dry cardboard can be ignited by a Raman laser. In fact, several solid materials can be ignited with a Raman laser, but liquids will not ignite. Liquids will convect and evaporate to dissipate heat before ignition occurs. Solids cannot convect and cannot dissipate heat as well as liquids. Cardboard will not ignite as long as it remains wet; however, if the laser dries an area of damp cardboard, the dry spot can ignite and spread to an area saturated by a flammable liquid. Scanning wet cardboard from a leaking package should not be done in bulk. It is much safer to cut away a small portion and scan it in a safe area.

Finally, be aware that the liquid on the cardboard may or may not have come from a leaking container within the box. Identifying a liquid is helpful, but be sure to confirm the finding by using other clues on-site.

Fluorescent Illicit Drugs

Some illicit drugs have low Raman response in relation to the high fluorescence produced. This makes it difficult for Raman instruments to identify heroin, cocaine and many other substances, especially when in lower concentrations after mixing with cutting agents.

Surface enhanced Raman spectroscopy (SERS) uses a solvent to suspend a target material in solution. A small metal wafer with specialized bonding surfaces is dipped into the solution to attract and hold the target material in place. The wafer is then air dried to remove the solvent, leaving trace amounts of target material on the wafer. (This extraction is effective regardless of fluorescent effect of the target material.)

The wafer/target complex is placed in the Raman laser and the signal is amplified exponentially to produce a Raman spectrum and identify the material (Figure 5.27).

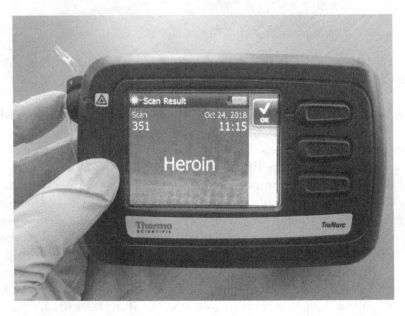

Figure 5.27 Surface Enhanced Raman Spectroscopy (SERS) is available for use with the Thermo Scientific TruNarc, a field-portable Raman instrument optimized to identify street drugs. In this image, the SERS wafer is being held in the laser focal point.

HPMS Techniques

A portable HPMS instrument can be used to identify trace amounts of airborne material and bulk or trace amounts of nonairborne material. However, it cannot be used for airborne and nonairborne materials simultaneously.

Airborne Material

An HPMS instrument continuously draws in and analyzes airborne samples. This means the instrument can identify material at a point source or to be mobile and used similarly to a traditional air monitor. Like an air monitor, the air drawn into the device is very close to the intake probe. The MX908 draws about 3 l per minute. As such, the device cannot "clear a room" or seek out a suspect chemical; it must be brought to the sample.

A methodical search pattern is helpful. Vapor density and air currents will affect sample concentration and dissipation. Consider sampling high, middle and low areas indoors with a perimeter search.

HPMS vapor analysis provides near real-time results for organic chemical vapors greater than about 50 g/mol. Total ion count is continuously updated and is roughly relative to concentration. A single HPMS instrument can be used in place of over 100 photoionization detectors, each set to a correction factor for over 100 unique chemicals.

Rule out strongly corrosive vapor or mist before exposing an HPMS or other electronic instrument to prevent damage, as HPMS cannot identify simple corrosives like mineral acids or ionic bases and damage may occur without warning.

Nonairborne Material

Solid material with low vapor pressure must be forced to become airborne so that the instrument can ionize it. This is accomplished when the operator collects a sample on a swab and then the instrument heats the swab and sample, causing vaporization.

Swab samples may be collected from bulk material, but care should be taken not to overload the swab. Very little material is necessary.

Figure 5.28 shows a testing sequence with an HPMS swab. A small cardboard package was suspected to contain an illicit drug and was intercepted in the Netherlands. The operator used a swab on the interior of the box and the outer surface of a toy stuffed bear. The swab was placed in the instrument, which thermally desorbed the sample material and then identified it as containing fentanyl. The toy bear was found to contain a hermetically sealed bag

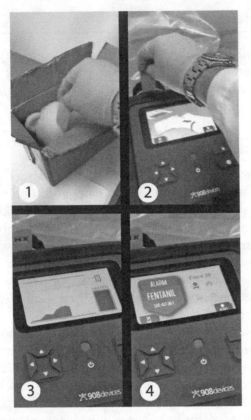

Figure 5.28 (1) Interior of box and contents are sampled with a swab. (2) Swab is placed in thermal desorption unit. (3) HPMS displays ion abundance and time remaining. (4) Alarm for fentanyl on display. Run time for the test was 60 sec. (Image: Kjellt Zomer, GATE Specialties, NL. With permission.)

of fentanyl. The swab collected trace amounts of fentanyl, presumably from cross-contamination during packaging.

A swab sample may be used to test several areas. For example, a single swab can be used on countertops, floor, handles and other items, even hands, when searching for a suspect chemical, such as an illicit drug. If the result is a chemical of interest, other swabs can be used to localize the target material. Otherwise, the operator can move on to another area.

The atmospheric ionization source causes MX908 to disregard common materials like sugars, starches, inorganic salts, hydrocarbons and so on. This helps the operator "sort" through dirty samples and quickly identify materials of interest.

HPMS instruments develop a mass spectrum, which is compared to a library of mass spectra. However, the spectrum is not viewable, unlike most Raman and FTIR instruments. A viewable mass spectrum would be of little use for orthogonal field testing since there is no spectral correlation across Raman, FTIR and mass spectrometry.

HPMS offers both high sensitivity and specificity, providing strong confidence in the identification of a relatively small, but important, group of chemicals. If HPMS is not producing an identification, an operator could expand sampling techniques after considering test specificity and other factors as described in Table 5.1.

Acid/Base Identification

Ionic acids and bases are difficult to identify with Raman, FTIR, and HPMS. Ionic bonds are too strong to respond, and dissociation leaves no bond to identify by Raman and FTIR. Materials with low molecular mass (<50 g/mol) and the inability to form ions in vapor phase prevent identification by HPMS. Complex anions may provide a signal but are difficult to discern from salts of the same anion. Raman will not identify mineral acids but may identify a complex anion. FTIR will not identify mineral acids or bases reliably across a range of concentrations due to the acid or base affecting the response from water. FTIR will identify acids and bases with complex ions; however, some simple acids will be compared to complex anionic acids such as sulfuric acid.

Some operators need to identify basic acids and bases while interrogating an illicit production process, as the process may or may not be dependent on a specific acid or base. Table 5.2 describes the response of common acids and bases that do not produce reliable Raman or FTIR response. The use of wide range pH paper-testing vapor and then liquid in combination with Raman

Table 5.1 Comparison of Test Properties

	Specificity	Sensitivity	Skill Level	Comments
pH Paper	High	High	Easy	
M8 Paper	Low	High	Easy	Aqueous inorganic liquids bead up, organics soak in
KI-Starch Paper	High	Medium	Easy	Some false-negatives
Colorimetric Tubes	High	High	Easy	Check cross-sensitivity
PID	Low	High	Moderate	Consider lamp ionization potential
IMS	Medium	High	Moderate	
HPMS	High	High	Easy	Limited library
GC/MS	High	High	Challenging	
Raman	High	Low	Easy	No trace detection
FTIR	High	Low	Moderate	No trace detection

Table 5.2 Simple Acid/Base Charateristics[1]

Corrosive	Liquid pH Result	Vapor pH Result[2]	Raman Signal	Characteristics FTIR Signal	HPMS Signal	Comment
Hydrochloric (muriatic) acid	Red	Probable	No	No	No	May cause white film on outside of container over time
Hydrofluoric acid	Red	Probable	No	No	No	Usually not stored in glass or metal unless plastic lined
Hydroiodic acid	Red	Probable	No	No	No	Clear or pale yellow liquid
Hydrobromic acid	Red	Probable	No	No	No	Colorless liquid in pure form. May take a yellow-brown tint
Sulfuric acid (if brown or black color)	Red	No vapor	Excess fluorescence	Poor	No	"Spent" sulfuric acid contains organic matter from a commercial process and is sold as drain cleaner
Nitric acid	Red	Probable	Strong	Strong	No	Extremely concentrated nitric acid will produce red fume. Raman may identify as a nitrate salt
Perchloric acid	Red	Probable	Strong	Strong	No	Raman may identify as a perchlorate salt
Ammonium hydroxide	Blue	Probable	No	Likely	No	Vapors lighter than air
Potassium hydroxide	Blue	No vapor	No	No	No	Also known as lye. May be used interchangeably with sodium hydroxide
Sodium hydroxide	Blue	No vapor	No	No	No	May be used interchangeably with potassium hydroxide

1 Other acids and bases are possible, but these are the most common that have no or poor Raman, FTIR, and HPMS response.

2 Strong volatile solutions will push corrosive vapor into the air just above the liquid, which is detectable, especially in higher humidity and temperature. A dry pH strip may indicate vapors in air above some corrosive or caustic liquids. Very dilute solutions may not produce enough vapor to produce a pH result. In arid conditions, moisten the pH strip before testing vapor.

and FTIR results may provide additional information on site. The comment section describes visual clues.

To use the table, record Raman, FTIR, and HPMS results. To perform a pH test, first hold the pH strip over the open container and watch for a color change. Volatile acids and bases will produce a result. Next, remove a drop by pipet (or other inert material), touch the pipet tip to the pH strip and record the results.

Working with Spectra

The purpose of this book is to enable field operators with knowledge and skills to be successful in characterizing or identifying unknown materials. Understanding some basic clues contained in your instrument generated spectra can help you to improve field tests and be more successful.

Raman Spectra

Modern Raman field instruments are designed to identify or classify a sample without user interpretation. Field operators are rarely spectroscopist and this is not an attempt to change the fact. However, a field operator may be able to recognize clues in a spectrum if the instrument returns an inconclusive result. For example, an operator's recognition of fluorescence or heat signatures may allow for sample manipulations to reduce the interference. Similarly, visual comparison of an inconclusive result with spectral overlays of suspected materials can lead to presumptive results.

When light strikes susceptible covalent chemical bonds, a Raman shift in light can be observed. This effect is more subtle than a reflection or fluorescence. The shift is specific to the bond but influenced by neighboring bonds. Each shift produced is represented by a peak in a spectrogram.

Spectrogram Basics

A Raman spectrum is the actual light energy induced by the instrument's laser. A spectrogram is the printed depiction of the light energy. What is expressed technically correct as a spectrogram is commonly called a spectrum by field operators. "Spectrum" will be used commonly in this publication.

An example of a Raman spectrogram is shown in Figure 5.29. The darker line depicts the spectrum from the operator's scan. The lighter line depicts a match to a spectrum stored in the instrument's database and labeled as ethylene glycol. The area below the lighter line has been filled in for clarity. The occurrence of a library match means the operator's scan met the criteria of the instrument that defines confidence in the result. Field instruments allow more mathematical flexibility in the analysis compared to laboratory

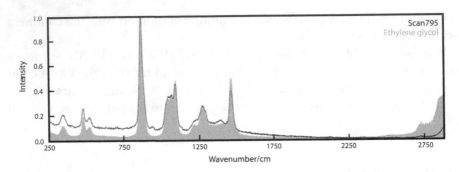

Figure 5.29 Example of a Raman spectrogram matched to ethylene glycol in the instrument database.

techniques. If the operator's scan did not produce a match, only the darker line would be displayed.

The vertical axis shows the intensity of the Raman-shifted light energy being collected by the instrument. It may show absolute values but is more likely to display a relative scale from 0 to 1.0, or 0% to 100%. Light energy developing 1% or 2% intensity can be produced by a low concentration or a weak Raman response. A high-intensity peak may be produced by a strong Raman response or somewhat higher concentrations. Light energy with an intensity above 1.0 (100%) will saturate the sensor and reduce signal fidelity.

The horizontal axis of a Raman spectrogram shows a wavenumber scale. Wavenumber is the number of waves per distance (centimeter), rather than the number of waves per time (second). In other words, frequency is the number of waves that pass by a fixed point in one second. Wavenumber is the number of waves in one centimeter. The unit may be described as "wavenumber/cm," which is the same as "wavenumber cm^{-1}."

The wavenumber axis is labeled with the range of wavenumbers detectable by the instrument. The range will vary by laser frequency and instrument capability. The example in Figure 5.29 is from an instrument with a 785 nm laser. The instrument is measuring light energy with a Raman shift away from the laser frequency, so laser energy would be at 0 on the wavenumber scale. Note in the example that wavenumber begins at 250. The range of 0–250 has been filtered out due to the high intensity of the laser compared to Raman-shifted light energy. Peaks produced with 0–250 wavenumbers are out of the range of detection for this instrument. The high end of the range tapers off around 2,800.

Raman instruments commonly using 532, 785, or 1064 nm lasers, as well as others, may be used to discover peaks that are out of range of a single instrument.

Fluorescence

Fluorescence in a Raman spectrum is caused by glare from the laser. Just as the oncoming headlights of a vehicle make it difficult for your eyes to discern nearby detail, so fluorescence blinds the instrument to nearby peaks in a spectrum.

As noted, low wavenumbers near the laser frequency are often not displayed because of constant fluorescence. Some substances will produce fluorescence in higher wavenumber ranges. Fluorescence is recognized in a spectrum as a sloping hill from left to right as shown in Figure 5.30. The broader and more intense the fluorescence, the more Raman-shifted peaks are likely to be buried within it.

Samples colored toward the green-blue-black end of the visible spectrum will produce more fluorescence than yellow-orange-red colors. On a grayscale of white to black, darker colors will produce relatively more fluorescence.

Fluorescence may be reduced by manipulating the sample in a way that reduces the intensity of the color. For example, a dyed oil may be heated in a test tube by torch until the dye degrades in the heat, leaving a clear liquid to test. A sample may be diluted with a solvent to dilute color intensity. Lowering laser intensity may decrease fluorescence.

Heat

The laser may induce heat in a sample. Mild heat production may produce enough infrared energy to blind the instrument. In this case, Raman-shifted peaks cannot be identified under the relatively more intense heat signature. High heat production may cause the sample to ignite or explode.

Heat may be recognized as a sloping line rising from the left, as a sawtoothed line or as a high line with peaks pointing downward (Figure 5.31).

A high-quality Raman spectrum shares two features: a low baseline and sharply defined peaks. A sloping hill, falling from left to right indicates

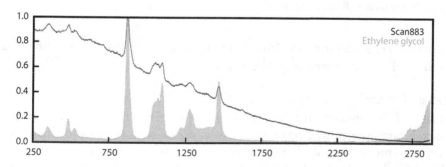

Figure 5.30 Example of moderate fluorescence in a Raman spectrogram. Peaks may be recognized within moderate fluorescence if the receiver is not saturated (>1.0). Weak Raman response may be muted by increasing fluorescence.

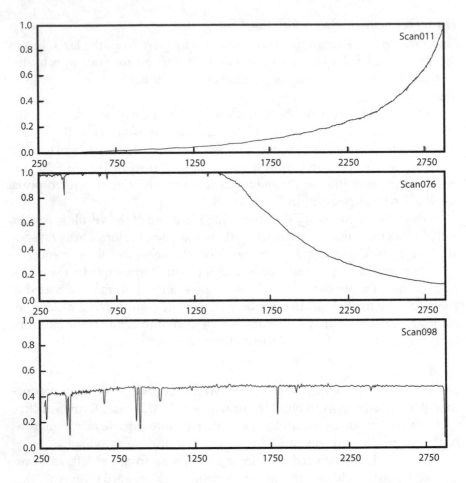

Figure 5.31 Three examples of Raman spectrograms influenced by heat. A sloping line from left to right (top), fluorescence with downward-pointing peaks (middle) and downward-pointing peaks without upward-pointing peaks (bottom) contain no useful Raman data.

fluorescence. A sloping hill rising from left to right, a saw-toothed line or a high baseline with inverted peaks all indicate heat.

Laser Power

Heat and fluorescence may be reduced by lowering the power level of the laser. Lower laser power may increase or decrease scan time. Some materials may need more scans at a lower power setting to produce an adequate Raman signal for the instrument. Other materials may require less scan time due to markedly decreased fluorescence, which allows an otherwise strong Raman response to shine through.

An alternative method is to slowly move the focal point over the surface of the sample material. This will reduce heat buildup in any one area of the sample, which will improve relative Raman response.

Figures 5.32–5.34 show three consecutive scans and results of palladium chloride on high, medium, and low laser power settings. Palladium chloride is a rich brown color, which should be suspected ahead of time to cause fluorescence and possibly heat. The palladium to chlorine bond is strongly Raman responsive.

The high setting, which most operators use as the default setting, incorrectly indicated a mixture of bromine and indium hydroxide as a group, mercury chloride and thallium chloride hydrate. Palladium chloride was not indicated.

The medium setting correctly indicated palladium chloride and incorrectly indicated a group containing bromine and indium chloride.

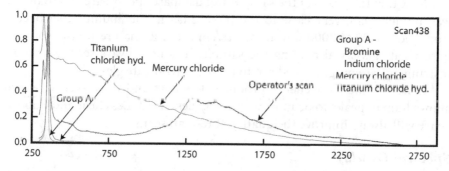

Figure 5.32 Raman scan with laser power set to high. The instrument incorrectly indicated a mixture of bromine and indium hydroxide as a group, mercury chloride and thallium chloride hydrate. Palladium chloride was not indicated. The wide wave in the operator's scan from approximately 1100 to 2000 is caused by the borosilicate glass vial and may be ignored.

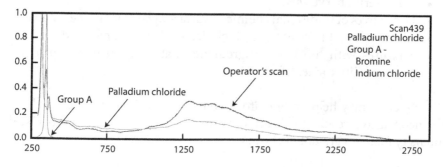

Figure 5.33 Raman scan with laser power set to medium. The instrument correctly indicated palladium chloride and incorrectly indicated a group containing bromine and indium chloride.

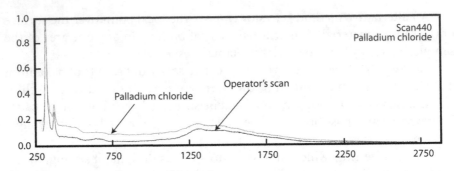

Figure 5.34 Raman scan with laser power set to low. The instrument accurately identified palladium chloride.

The low setting accurately identified palladium chloride.

Note that the peaks of the sample and database spectra are less than 400 wavenumbers/cm, very close to the intense laser. The "bump" that occurs from about 1100 to 2000 wavenumbers/cm is the Raman response from the borosilicate glass vial holding the sample. It is not always seen, and most instruments are programmed not to report borosilicate glass.

Lowering the laser energy produces less heat and fluorescence, which allows Raman peaks to be more accurately described. Less heat and fluorescence will always improve the quality of Raman scans.

Spectral Overlay

If the instrument has an overlay feature, an operator can choose a spectrum in the database and place it over a spectrum from the sample. This may be useful in the case of:

- A novel fentanyl derivative
- An A-series nerve agent
- Derivatives of traditional chemical warfare agents not in the database
- Chemicals that have not been included in the instrument's database
- Mixtures with too many ingredients that overwhelm the instrument's ability to identify

On-site clues may help an operator determine which spectra to overlay on an inconclusive result. Look for peak matches on the wavenumber scale. Intensity matches are not necessary.

FTIR Spectra

Modern FTIR field instruments are designed to identify or classify a sample. Without the qualifications of a spectroscopist, a field operator may be able to recognize clues in a spectrum if the instrument returns an inconclusive result. For example, an operator's recognition of an overwhelming water signature may allow for sample manipulations to reduce the interference. Similarly, visual comparison of an inconclusive result with spectral overlays of suspected materials can lead to presumptive results.

Spectrogram Basics

At first glance, an FTIR spectrum seems similar to a Raman spectrum, but they are distinctly different. An FTIR spectrum is the remainder of a broad band of infrared light after absorption by the sample.

The process starts with a background scan, which defines a band of infrared light. The known band of light is then projected on the sample, which absorbs frequencies of light characteristic of individual covalent bonds. The instrument records the reflected light minus the absorbed frequencies, which is missing from the known background scan. The spectrum of returning infrared light is absent light energy as shown in Figure 5.35. (A spectrogram is the printed depiction of the light energy. What is expressed technically

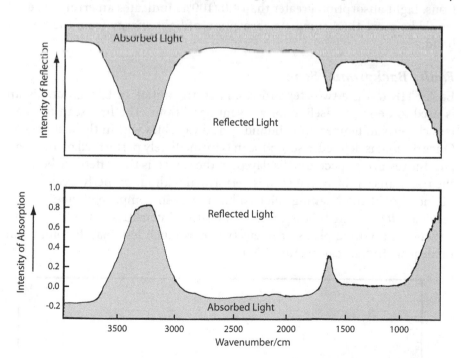

Figure 5.35 A band of infrared light illuminates a sample, and characteristic covalent bonds absorb portions of light energy. The instrument receiver detects reflected light energy minus the absorbed portions (top). The spectrogram is inverted to represent the absorbed infrared light as peaks and waves (bottom). The intensity of the absorption is measured on a scale of 0–1.0 (0%–100%); a value of 1.0 means no light energy was reflected into the receiver.

correct as a spectrogram is commonly called a "spectrum" by field operators and will be used commonly in this publication.)

The spectrum is inverted to show the areas of absorption as peaks, thus denoting areas of interest (Figure 5.36).

The darker line depicts the spectrum from the operator's scan. The lighter line depicts a match to a spectrum stored in the instrument's database and labeled as ethylene glycol. The area below the lighter line has been filled in for clarity. The occurrence of a library match means the operator's scan met the criteria of the instrument that defines confidence in the result. Field instruments allow more mathematical flexibility in the analysis compared to laboratory techniques. If the operator's scan did not produce a match, only the darker line would be displayed.

The vertical axis shows intensity of absorbed light energy. It may show absolute values but is more likely to display a relative scale from 0 to 1.0, or 0% to 100%. Light energy developing 1% or 2% intensity can be produced by a low concentration or a weak FTIR response. A high-intensity peak may be produced by a strong FTIR response or somewhat higher concentrations. Light absorption greater than 1.0 (100%) indicates an error, since 1.0 would be more than complete absorption of light energy from the known band.

Faulty Background Scan

Each FTIR test is a two-step process. First, the initial, or background, scan is used to define a baseline for ambient conditions. The first scan normalizes changes in temperature, humidity and variables within the instrument. Once normal is defined, a second scan is immediately performed on the sample. The resulting spectrum displayed on the screen is the difference between the two scans. If the initial scan is performed without properly cleaning the diamond or if the cleaning solution has not been completely removed, the residual effects will be displayed as negative intensities. In other words, downward-pointing peaks or intensity values much less than 0 point to an inadequate initial scan (Figure 5.37).

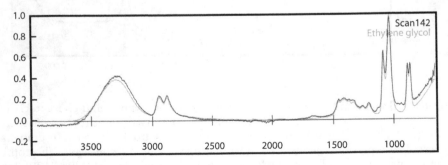

Figure 5.36 FTIR spectrum ethylene glycol.

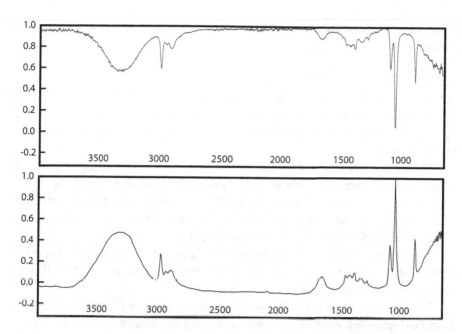

Figure 5.37 An FTIR scan after an initial scan (top) performed with residual ethanol/water cleaning solution on the diamond. Compare to a proper scan of the same ethanol/water solution. Any residual material left on the diamond surface will display as a negative value since the background scan defines a baseline before a sample material is tested.

In practice, the effects of a dirty diamond are observed as downward-pointing peaks interspersed with upward-pointing peaks from the sample material. Scan results with downward-pointing peaks should be repeated even if the instrument delivers and identification because the negative peaks may be muting the presence of other mixture components.

Vibration

Most FTIR instruments use a moving mirror to split the infrared beam. The mirror must move uniformly along its path to record peaks in the correct locations. Vibration or any sudden movement can momentarily interrupt the movement and inject what appear to be peaks into the spectrum. Figure 5.38 shows evidence of vibration to the extent that the instrument was not able to identify the sample despite the presence of several taller peaks.

Vibrational interference may be caused by:

- Moving the instrument during a background scan or sample scan
- Bumping the table on which the instrument is resting
- Vehicle engine vibration or vehicle movement

If vibrational interference is observed, simply repeat the test in a quieter environment.

Water

Water is strongly FTIR responsive and can mute less responsive materials. This means lower concentrations of materials in water may not be identified. Pairing a Raman test with an FTIR test significantly improves accuracy on random samples.

Figure 5.39 compares Raman and FTIR scans on 3% hydrogen peroxide in water. Hydrogen peroxide is FTIR responsive and shows a bump in the 3000-wavenumber area. The remaining 97% of the solution, water, is strongly FTIR responsive and overwhelms the weak peroxide signal. The instrument correctly identifies water yet misses hydrogen peroxide.

Water has a Raman response, but it is generally outside the window of identification for field instruments. Hydrogen peroxide is strongly Raman responsive, displaying a bold, single peak.

Pairing Raman and FTIR testing increases accuracy when testing random field samples, especially when dark samples induce heating in Raman testing. FTIR is not impeded by color and does not induce heat.

Anecdotal reports indicate random field test accuracy improves approximately 30% when Raman and FTIR tests are used in a complementary manner over individual use.

Note the similarities and differences of Raman and FTIR spectrograms in Figure 5.39. The y-axis describes intensity of Raman-shifted light or absorbed infrared light; the two are not comparable. The x-axis shows activity across wavenumbers. Raman wavenumbers increase from left to right while FTIR wavenumbers decrease from left to right. The Raman spectrum shows evidence of fluorescence and borosilicate glass vial with a peroxide peak and no evidence of water. The FTIR spectrum shows evidence of water only (the peroxide peak is obscured in the 3000-wavenumber area) and is not affected by fluorescence or a glass vial.

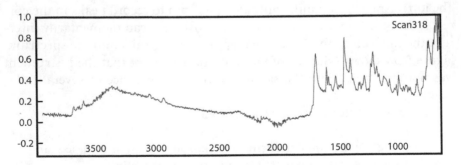

Figure 5.38 Vibration is evident in this FTIR spectrum in the areas of 4000–3100 and 2300–1800.

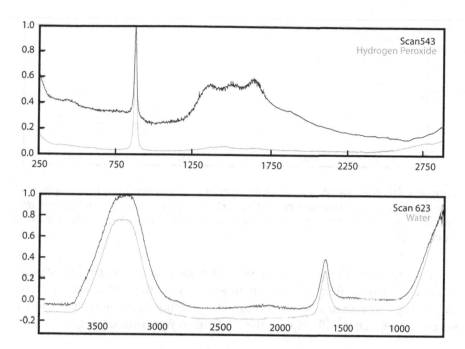

Figure 5.39 Raman (top) and FTIR (bottom) spectrograms of 3% hydrogen peroxide in water. The Raman instrument does not identify water and only identifies hydrogen peroxide. The FTIR instrument identified water but was not able to identify the low concentration of hydrogen peroxide, which produces a small shoulder, at about 3000 wavenumber, on the water peak.

Spectral Overlay

Much like other instruments with an overlay feature, an operator can choose a spectrum in the database and place it over a spectrum from the sample. This may be useful when the instrument displays a spectrum with strong characteristics but does not produce an identification, for example:

- A novel fentanyl derivative
- An A-series nerve agent
- Derivatives of traditional chemical warfare agents not in the database
- Chemicals that have not been included in the instrument's database
- Mixtures with too many ingredients that overwhelm the instrument's ability to identify

On-site clues may help an operator determine which spectra to overlay on an inconclusive result. Look for peak matches on the wavenumber scale. Peak intensity not as important as peak location (wavenumber).

Mixtures can cause stress on individual bonds causing peaks to shift from expected locations. Differing concentrations can increase or decrease the displayed intensity of peaks (Figure 5.40).

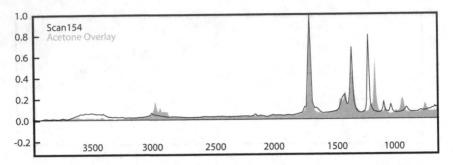

Figure 5.40 Acetone (shaded peaks) compared to operator's scan. The operator's scan (dark line) is a mixture of methyl ethyl ketone, acetone, tetrahydrofuran, cyclohexanone and water absorbed from the air. Shifted peaks are apparent with an overlay.

An operator should perform several tests before resorting to spectral analysis. Consider altering instrument settings, manipulating sample material, using more than one technology and paying meticulous attention to detail. These tactics will help get the most from your instrument.

Bibliography

Bennett, William, *Personal Notes*, Galesburg, MI, 2017–2020.
Brown, Christopher, *Personal Notes*, 908 Devices, Boston, MA, 2020.
ChemThief Owner's Manual, Version 1.1, Ideal Products, Nicholasville, KY, 2019.
Gemini User Guide, 110-00107-06, Thermo Scientific, Tewksbury, MA, 2018.
Houghton, Rick, *Personal Notes*, St. Johns, MI, 2008–2019.
Jaroh, Adam, *Personal Notes*, Telerob USA Inc., Erie, PA, 2020.
TruNarc User Guide, 110-00056-03, Thermo Scientific, Tewksbury, MA, 2018.

Index

Printed in the United States
By Bookmasters